Spaceflight
Revolution

Spaceflight Revolution

David Ashford

Bristol Spaceplanes Limited, UK

Imperial College Press

Published by

Imperial College Press
57 Shelton Street
Covent Garden
London WC2H 9HE

Distributed by

World Scientific Publishing Co. Pte. Ltd.
5 Toh Tuck Link, Singapore 596224
USA office: Suite 202, 1060 Main Street, River Edge, NJ 07661
UK office: 57 Shelton Street, Covent Garden, London WC2H 9HE

British Library Cataloguing-in-Publication Data
A catalogue record for this book is available from the British Library.

SPACEFLIGHT REVOLUTION

ISBN 1-86094-320-9
ISBN 1-86094-325-X (pbk)

Typeset by Rocket Media, Flax Bourton, Bristol, England. Info@rocket-media.co.uk

Printed by FuIsland Offset Printing (S) Pte Ltd, Singapore

CONTENTS

PREFACE

The genesis of this book goes back to a spring day in 1960 during my final year as an aeronautical engineering student in London at Imperial College. I was being interviewed for a job with the Hawker Siddeley Aviation Advanced Projects Group. The conversation went something like this:

"How did you come to this interview, my boy?"

"I cycled, sir".

"I see. How much cycling would you do if your bike were scrapped after each ride?"

"Er, not much, sir".

"Quite right, my boy. Well, that is how it is with launching satellites. They are carried on converted ballistic missiles that can fly only once. Very expensive. We are going to add wings, a tail, a cockpit, a heat shield and landing gear, and fly them back. All you have to do is refuel them and fly them again, just like an aeroplane. Greatly reduced cost. Greatly improved reliability. Obvious, isn't it?"

I must have looked duly impressed by the power of this logic, because they offered me a job as an aerodynamicist in the hypersonics team. I accepted this offer and started in 1961 after a year at Princeton doing post-graduate research on rocket motors. I would probably have taken another job if I had known that, forty-two years later, all satellites would still be launched by largely expendable vehicles, and that I would be writing a second book trying to revive this 1960s dream. (The first, *Your Spaceflight Manual - How You Could be a Tourist in Space Within Twenty Years*, was co-written with Dr Patrick Collins and published by Headline in 1990.)

For a few years, the job was as interesting as the interviewer (John Allen, who still remembers his patter) had promised. We were planning aeroplanes that could fly to and from orbit to replace the ballistic missiles then being used to launch satellites, as well as airliners two or three times faster than Concorde and sub-orbital airliners that could fly from New York to Tokyo in 75 minutes. We were planning a great advance in aeronautics and astronautics. Most large aerospace companies at the time had comparable ideas. In the United States, the X-15 research aeroplane was demonstrating the required technology by flying regularly to a height greater than 50 miles (80 km) which is usually considered to be the lower limit of space.

This great advance has never happened.

Space policy was diverted by the pressures of the Cold War into using throwaway vehicles for carrying people to space, and it is still stuck in the same rut. Even the Space Shuttle failed to meet its promise of reducing the cost of access to space because it was built far more like a ballistic missile than an aeroplane. The X-15 remains the only fully reusable vehicle to have been to space and back.

The present way of doing business is rapidly becoming untenable, and a return to the plans of the 1960s is likely soon. A rapid catching up is possible, using technology since developed for other projects. Among the benefits of the spaceflight revolution would be that in about ten years from now most people who are reasonably fit and affluent will be able to take a brief trip to space. The book explains what could have happened, why it did not happen, and how and why it is likely to happen soon. The predictions are based on straightforward and robust assumptions and analysis, using simple first-order sizing and costing comparisons. The derivation of some of the results is provided in the appendices, for those who wish to delve further.

I should declare a commercial interest in that I have kept an involvement with launch vehicles and now run a small company that has plans to develop a small spaceplane. I have done my best to prevent this from biasing the argument. If anyone feels that the result is unfair, by commission or omission, I hope they will accept that this is more of an exposition than a source book. They are most welcome to comment. An open debate on the way ahead for space would benefit us all.

I am greatly indebted to friends and colleagues in the small but rapidly growing space tourism movement for their help and support over the years, and hope that they in turn will find this book helpful.

I would like to thank all those who have helped with this book, especially Mark Radice for much of the research, Professor Mark Birkinshaw for advice on space-based astronomy, Dr Patrick Magee for advice on medical aspects of spaceflight, Professor Ray Turner for advice on radiation in orbit; and Dr Robert Hall and Carol Pinchefsky for very helpful comments on the draft. Any misinterpretations of this advice are entirely my responsibility.

I would also like to thank various copyright owners for permission to use their illustrations.

LIST OF FIGURES

GLOSSARY

Ballistic Missile Used here in the modern sense to refer to large rocket-powered guided weapons with no wings.

Carrier Aeroplane The aeroplane-like lower stage of a multi-stage launch vehicle.

ELV Expendable Launch Vehicle.

Fuel The chemical substance that reacts in an engine with the oxidiser. In a rocket-powered aeroplane, 'Fuel' sometimes used loosely to mean fuel and oxidiser, i.e., all the propellant.

g Used here to mean a person's apparent weight divided by that at the Earth's surface. A person at the Earth's surface supported by a chair, floor, etc., will 'feel' 1g. This feel is due to the restraint preventing him/her from falling towards the centre of the Earth, pulled bygravity, at an acceleration of approximately 9.8 metres per second per second. In orbit in a non-rotating spacecraft, there are no restraining forces, and the person will feel weightless, or experience 'zero-g'.

Hypersonic Speeds faster than about Mach 5, or five times the speed of sound. At this and higher speeds, the shock waves are close to the aircraft surface, and aerodynamic characteristics do not change greatly between Mach 5 and orbital velocity, which is approximately Mach 25.

Ideal Mature Spaceplane Used here to mean the most competitive spaceplane design which could be built using technology at present at the stage of early demonstration.

Inert Mass The take-off mass of an aeroplane or launcher can conveniently be divided into inert mass, propellant mass, and payload. The inert mass consists mainly of structure, engines, and equipment. In an expendable launcher, the heaviest structural items are the propellant tanks.

ISS International Space Station.

Mach Number Aircraft speed divided by the local speed of sound, which is about 340 m/s at sea level and 295 m/s in the stratosphere, depending on local temperature.

Mature A new technology starts with research and then evolves through demonstration, prototype construction and early operations to mature operations. Cost and ease of use greatly improve during this maturing process. A mature spaceplane is defined here asone which is as good as an airliner in terms of safety, life, maintenance cost, and turnaround time.

Operational Prototype Used here to mean an early development model of a new flying machine that is built in an experimental workshop and used for flight testing and some early operational missions.

Orbit A trajectory of a small body in space around a large one which is periodically repeated, such as a satellite around the Earth. Unless stated otherwise, refers to Low Earth Orbit, i.e., a more or less circular orbit round the Earth at a height of a few hundred kilometres.

Orbital Velocity The horizontal speed needed for a body to circle the Earth in Low Earth Orbit. Approximately 8 km/sec. 'Velocity' used when direction is relevant, 'speed' when it is not.

Orbiter The upper stage of a multi-stage reusable launch vehicle that reaches orbit.

Oxidiser The chemical substance consumed in an engine that reacts with the fuel. In a jet engine, the oxidiser is the oxygen in the air. In a rocket engine, the oxidiser is carried on board the vehicle in a separate tank. It is often liquid oxygen or a chemical containing oxygen.

Propellant Fuel and oxidiser carried in a vehicle for use in a rocket engine.

RLV Reusable Launch Vehicle.

Rocket Engine An engine in which fuel and oxidiser, both carried by the vehicle, combine chemically in a combustion chamber to generate a high temperature gas, which exits through a nozzle to produce thrust. 'Rocket' often used loosely to describe a complete rocket-powered vehicle that takes off vertically and is expendable, like a rocket firework, ballistic missile, or conventional launch vehicle.

Space Height As height above the Earth increases, the atmosphere becomes less dense. Space height is the height at which the atmosphere is deemed to stop and space to start. This is largely a matter of definition but, for aeronautical engineers, it is the height at which the air loads on a vehicle become very low. There are various definitions of space height. In the 1960s, the USAF awarded Astronaut Wings to X-15 pilots who exceeded a height of 50 miles (80 km). At this height, the sky is dark with bright stars even in daytime. At this height, air loads become significant on the Space Shuttle as it reenters the atmosphere from orbit. The Federation Aéronautique Internationale defines the edge of space as 100 km (62 miles). The FAI is the international body regulating aircraft speed and other records. The 'A stands for Aéronautique rather than Astronautique, and its jurisdiction stops at 100 km. For present purposes, the 50 mile (80 km) definition is convenient. This is about seven times higher than the cruise height of jet airliners.

Spaceplane Used here to mean an aeroplane-like fully reusable vehicle that can fly to and from space height and that lands using wings for lift. Unless prefixed by 'sub-orbital', refers to a vehicle that can fly to and from orbit.

Spaceplane Age Used here to refer to the time in the future when spaceplanes have revolutionised space travel, much as the railways did to land travel.

Specific Impulse A measure of rocket motor efficiency. The thrust produced for each kg of propellant consumed per second.

Stage Launchers are an assembly of vehicle elements, all but one of which are discarded during the acceleration to orbital velocity. These elements are called stages. Each stage increases the velocity until orbit is achieved. A single-stage vehicle is theoretically possible, but has not yet been built.

Sub-orbital A trajectory that reaches space height, but not sufficient speed to stay up like a satellite. Unless prefixed by 'long range', refers to a vehicle that spends only a few minutes in space and lands back at or near to where it took off.

Sub-orbital Shuttle Aeroplane A 'thought experiment' modification of the Space Shuttle Orbiter, to enable it to take off down a runway unassisted by expendable components.

Turnaround Time The time required at an airport terminal to unload an airliner, prepare it for the next flight, and re-load it.

Zero-g A convenient term to describe the weightless conditions experienced in orbit in a non-rotating spacecraft. Strictly speaking, the 'g' is nearly always more than zero, and 'microgravity' is a more precise term.

Zoom-climb A flight manoeuvre in which speed is traded for height. The aeroplane starts at high speed in level flight, the pilot pulls it into a steep climb, and the speed drops off as height is gained. Zoom-climbs are a means of briefly reaching great height.

CHAPTER 1 - INTRODUCTION

This book describes a straightforward analysis showing that a revolution in spaceflight is likely to start soon, which would lead to a thousandfold reduction in the cost of sending people to space within about fifteen years. Within this timescale, short visits to orbiting hotels by fare-paying passengers are likely to become the largest business in space. This great reduction in launch cost would be achieved at far less cost to the taxpayer and far sooner than by present official plans. The main requirement is for commerce to replace politics as the driving force behind space policy.

The key to the revolution is to replace throwaway launchers like ballistic missiles with aeroplanes capable of flying to space, and then to develop these spaceplanes to airliner standards of safety, life, maintenance cost and turnaround time. Ballistic missiles can fly only once. Such expendability is unique among routine transportation systems and is the fundamental cause of the high cost and risk of spaceflight. Imagine how much motoring would cost if automobiles could be used for only one journey!

Part 1 of the book describes the past, present, and predicted future of spaceplanes. Within this Part, Chapter 2 describes how spaceplanes first became feasible in the 1960s but were not developed because of the politics behind the NASA budget. Chapter 3 describes recent developments, and Chapter 4 summarises a predicted way ahead for spaceplanes based on some of the more promising early projects and ideas developed since then. The rest of the book describes the analysis to support these predictions.

Part 2 considers the potential of the spaceplane space age, showing how mature spaceplanes will reduce launch costs by a factor of 1000, making possible large new commercial industries in space, of which space tourism is likely to be the largest. Chapter 7 includes a description of a visit to a space hotel, the highlights of which will be superb views of the Earth, and fascinating activities under weightless conditions.

Most space professionals would probably agree that this potential will be achieved some day. The consensus in government space agencies, to the extent that there is one, is that it will only be realised some forty to fifty years from now. Part 3 provides the supporting evidence behind the prediction that it will happen in about fifteen years. It considers technical feasibility, safety, the developments needed for technical maturity (and affordable space tourism), the size of the market, the reasons why a fully reusable launch vehicle can be developed at less cost than an expendable one, a design logic for the new spaceplanes, and what it will take to break the mould of traditional thinking on space policy.

The final two chapters consider the benefits of low-cost access to space, and summarise the conclusions.

The main conclusion is that the pivotal next step is the development of a small aeroplane that can climb briefly to space height - a sort of re-invented X-15, which flew successfully to space during the 1960s. (The X-15 is mentioned many times in this book, as it was the first to demonstrate the feasibility of aeroplanes capable of flying to and from space). This would build up the market for spaceplanes and provide a focus for maturing the technology. By flying to space several times per day, it would demonstrate the potential of spaceplanes and break the mould of traditional thinking about space transportation. All this could be achieved at the cost of three or four fighter planes off the production line.

This is one of the very rare situations in which something that appears to be too good to be true actually is true. The general perception is that space transportation has to be difficult and expensive, and that present government space policy is sound. This book concludes that the reality is the exact opposite - affordable space travel is well within our reach.

PART 1

Past, Present, Predicted Future

CHAPTER 2 - SPACEPLANE HISTORY

A great opportunity was missed in the 1960s to transform space transportation, and we are still living with the consequences. All satellites had been launched by ballistic missiles or developments thereof, which suffered from the fundamental disadvantage of being able to fly only once. Most large aerospace companies had teams studying reusable launch vehicles and there was a consensus that these were the obvious next major advance, offering greatly reduced launch cost and improved reliability. In the United States, the X-15 research aeroplane was flying regularly to space height (50 miles, 80 km) and demonstrating much of the required technology.

Yet, the X-15 remains the only fully reusable vehicle to have been to space - which it last did in 1968 - and reusable launch vehicles have never been developed. The explanation lies in the history of launcher development in the United States, which has been the leading space power for most of the space age.

Figure 2.1 The V-2 Ballistic Missile of World War II [Deutsches Museum]
The V-2 was the first vehicle to reach space height, and was the progenitor of all subsequent ballistic missiles and launch vehicles.

Significant U.S. work on ballistic missiles started with German V-2s, Figure 2.1, captured at the end of World War II. In 1942, the V-2 had become the first vehicle to reach space height. It was used mainly against London in the last year of World War II.

The Allies were very impressed by the engineering of the V-2, which was far ahead of anything they had, even on the drawing board. They were less impressed by its effectiveness as a weapon. Its miss distance was such that the only practical target was a large city, and it was unlikely to damage more than a few buildings. It probably cost more to build than the material value of the damage it caused.

The advent of nuclear warheads changed this perception of ballistic missiles, as vast damage could now be caused. A deadly race then began between the United States and the Soviet Union to develop a nuclear-armed intercontinental ballistic missile. These became operational in the early 1960s, and both powers could then launch devastating attacks from their home territory. Capable of flying one third of the way around the Earth in less than half an hour, they are still the most destructive weapons the world has ever seen and the hardest to defend against.

Since ballistic missiles can fly to space, it was natural to use them to launch the first satellites. Figure 2.2 shows the subsequent evolution of U.S. launch vehicles.

Many key German engineers went to the United States at the end of World War II. The best known of these was Wernher von Braun, who was a leading pioneer of German rocketry in general and the V-2 in particular.

The first indigenous U.S. ballistic missile was the Redstone, developed by von Braun's team for the Army. Redstone was in effect a re-engineered and enlarged V-2. It served as the first operational U.S. ballistic missile and as the lower stage of the Juno that launched the first U.S. satellite in 1958, in rapid response to the brilliant and unexpected Soviet achievement of launching the first satellite, Sputnik, in 1957.

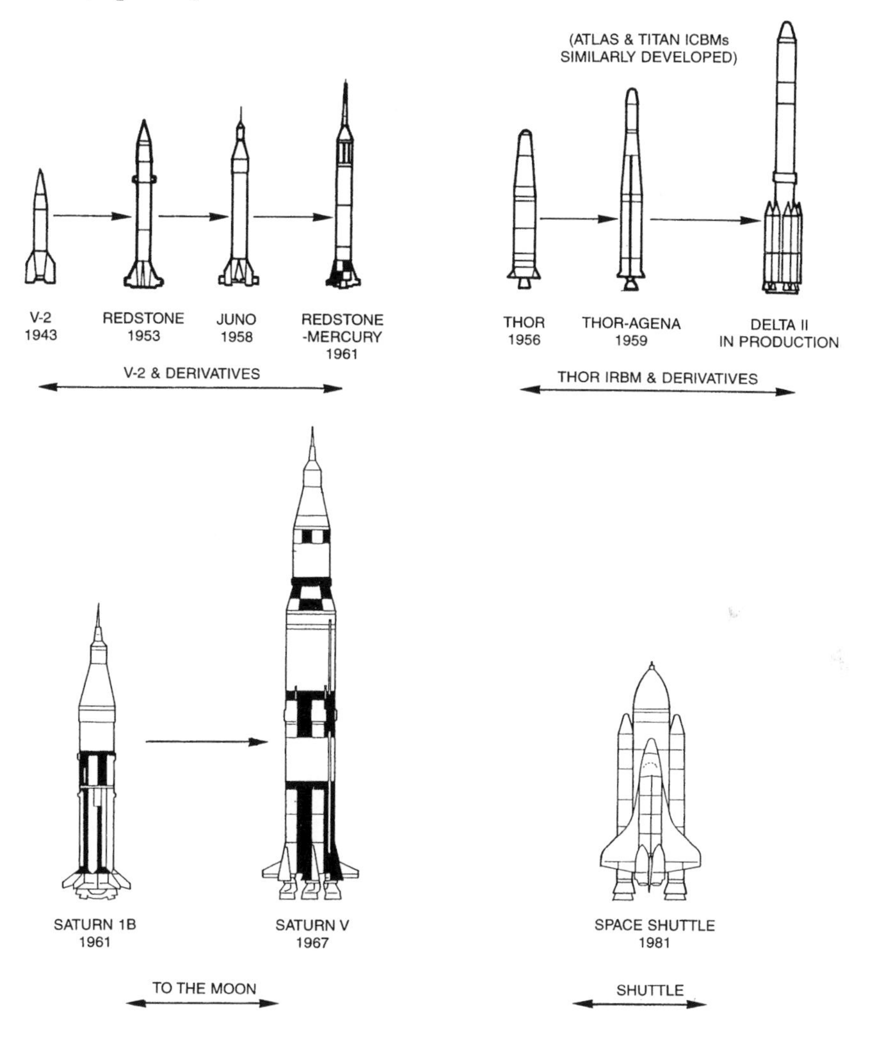

Figure 2.2 U.S. Launch Vehicle Evolution

The U.S. space programme evolved over thirty-five years from captured German V-2 ballistic missiles to the Space Shuttle. All these vehicles are expendable, except for the Space Shuttle, which is partly reusable.

Redstone also served as the launcher for the Mercury capsule in which Commander Alan Shepard, USN, became the first American in space in May

1961, albeit on a sub-orbital trajectory. This was again in rapid response to another major Soviet achievement - the successful launch and recovery of the first person to space, Yuri Gagarin, in April 1961. The consensus in the 1950s had been that the first people to space would get there in rocket-powered aeroplanes. It was the pressures of the Cold War that led to ballistic missiles being used.

Redstone was followed by the Jupiter and Thor intermediate-range ballistic missiles, and then by the Atlas and Titan intercontinental ballistic missiles. Titan was the last of these to fly, which it did for the first time in 1959. All four were also used as the lower stage of launch vehicles. None are still used as missiles, but derivatives of the latter three are still in production as launchers (Thor is now called Delta, after an earlier upper stage), although the latest versions have little in common except for the name.

The atmosphere in the United States at the time was of intense fear about the capabilities and intentions of the Soviet Union, as very well described in Tom Wolfe's *The Right Stuff* (New York: Farrar, Straus, Giroux, 1979). He explains why the early U.S. astronauts were acclaimed as national heroes to an extent that surprised even NASA, which had been formed in 1958, enabling them to make bold plans with big budgets.

In May 1961, just six weeks after Gagarin's flight, President Kennedy made the historic speech that galvanised the nation: "If we are to win the battle that is going on around the world between freedom and tyranny, if we are to win the battle for men's minds... I believe that this nation should commit itself to achieving the goal, before this decade is out, of landing a man on the Moon and returning him safely to Earth."

Thus began the race to the Moon. The first sentence in the quotation shows how political the objectives were. Project Apollo was perhaps the greatest engineering achievement of all time. Twelve men stood on the Moon between 1969 and 1972, and all got back safely. The Soviet effort was an expensive failure. The West went on to win the Cold War, Apollo having played its part.

Von Braun was responsible for the mighty Saturn launchers used in the Apollo programme. His original ideas for a large launcher, published in 1952 [20], were for a fully reusuable vehicle using experience from the winged V-2 (see later). However, reusability would have delayed Apollo and would not have saved much money because of the small number of launches. Saturn was therefore expendable, like a greatly enlarged ballistic missile.

NASA followed Apollo with the Skylab space station and with the Space Shuttle. Skylab, Figure 2.3, was the first U.S. space station. The main structure was converted from the third stage of a Saturn 5 launcher and had a useable volume equivalent to that of two railway carriages. It was launched on top of a Saturn in 1973, and three teams of astronauts each spent up to three months on board in 1973 and 1974, carrying out valuable scientific research and solving many of the problems of living and working in space.

Figure 2.3 The Skylab Space Station [NASA]
Skylab was the first U.S. space station, and operated very successfully during 1973 and 1974.

The original plan was for the Space Shuttle to be fully reusable. The technology for a fully reusable vehicle was certainly available by this time, and we will discuss this point before returning to the decision to build a shuttle that was not fully reusable, shown in Figure 2.4.

Figure 2.4 The Space Shuttle [NASA]
The Orbiter is reusable, but the large External Tank burns up on re-entry, and the two Solid Rocket Boosters are recovered at sea and re-cycled. The Shuttle first flew in 1981, and has averaged five flights per year since then at a cost of some $1 billion per flight.

In parallel with the development of ballistic missiles and related launchers, there was a major research effort on high-speed rocket aeroplanes. The first to fly were two German V-2s fitted experimentally with wings, shown in Figure 2.5. The aim was to increase range by means of a gliding re-entry instead of the ballistic one, so that missiles could be launched out of range of Allied bombers.

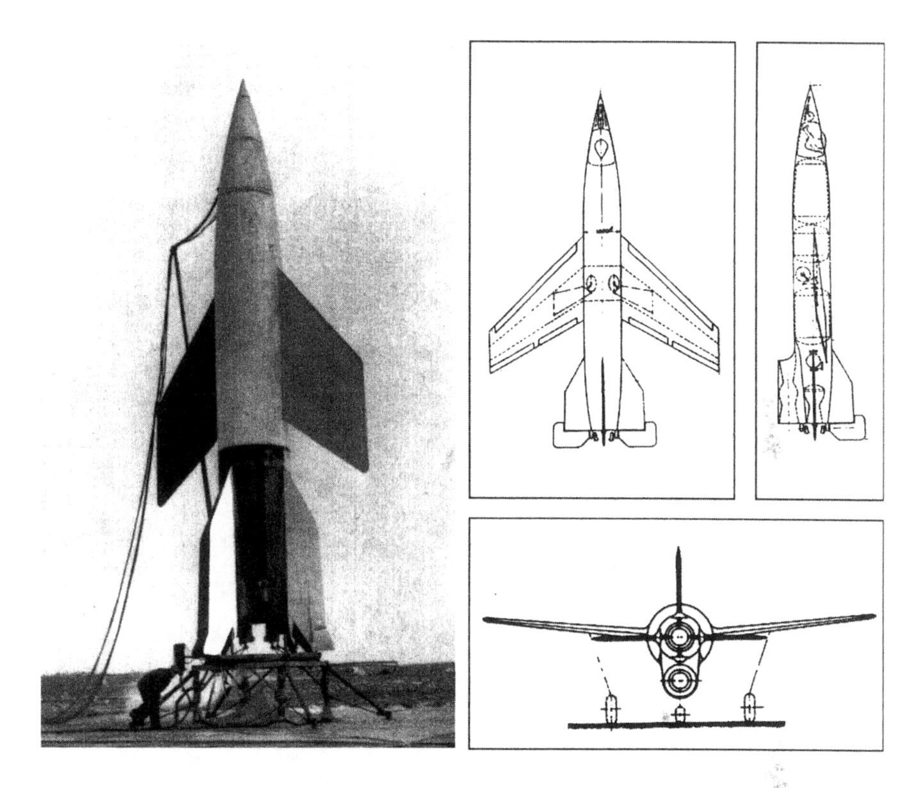

Figure 2.5 Winged Developments of the V-2 [Deutsches Museum]
The experimental winged V-2 (left) was the first full-size supersonic aeroplane, reaching Mach 4 in 1945. A piloted version (right) was considered but not built. It could have served as a reusable launcher lower stage.

The first winged V-2 blew up shortly after launch for reasons unconnected with the wings. The second reached Mach 4 in stable and controlled flight in the atmosphere in January 1945. One wing came off on re-entry, probably due to excessive air loads. Nevertheless, this was a remarkable achievement. It would be sixteen years before a full-size aeroplane flew faster (the X-15).

There was also a project for a piloted version of the winged V-2. This would have taken off vertically, ascended to space height, reentered the atmosphere, glided to the target, dropped a bomb, and flown back using a ramjet engine, ready for the next mission. This project was probably not much more than a sketch but even so, given priority, it could probably have been built in a year or two.

It is relevant to speculate on the consequences if this project had been developed and if the United States had captured piloted V-2s rather than unpiloted ones. The piloted winged V-2 would have been essentially an aeroplane, comparable in size and performance to the X-15. It would have been flown by aeroplane pilots, serviced by aeroplane mechanics, and had in-service modifications to improve performance, reliability and safety like any other operational aeroplane. The design team would have included aeroplane specialists. In short, an aeroplane culture would have taken over from a missile one, driven by the requirement to provide pilot safety up to contemporary military standards. This culture would have spread to the early space programme. Vehicles with a high-speed, piloted, reusable lower stage would probably have launched the first satellites. We are still waiting for such a development.

Shortly after World War II, the United States embarked on a programme of high-speed rocket research aeroplanes, Figure 2.6.

The Bell X-1 was the first piloted aeroplane to fly faster than sound, i.e., to break the so-called sound barrier or to exceed Mach 1 (1947); the Douglas Skyrocket was the first to Mach 2 (1953); the Bell X-2 was the first to Mach 3 (1956); and the North American X-15 was the first to Machs 4, 5, and 6, (all in 1961).

The aim of the X-15 was to develop the technologies for high-speed, high-altitude flight, for use in the space programme and reusable launch vehicles. It made 199 flights between 1959 and 1968, of which 13 exceeded a height of 50 miles (80 km), entitling their pilots to USAF astronaut wings. The first of these was USAF Major Robert White who, on 17 July 1962, became the first person to fly to space and subsequently land back at his point of origin.

Bell X-1A.
Developed from X-1, the first supersonic aeroplane, 1947.

Douglas Skyrocket,
launched from B-29.
First to Mach 2, 1953

Bell X-2.
First to Mach 3, 1956

North American X-15
Fastest and highest true aeroplane.
1961

Figure 2.6 U.S. Rocket-Powered Research Aeroplanes [NASA]
This series of high-speed rocket aeroplanes had paved the way for a reusable launch vehicle by the early 1960s.

The X-15 reached a maximum height of 108 km (six times higher than Concorde) and a maximum speed of Mach 6.7 (more than three times faster than Concorde). There was a proposal in 1961 for the X-15 to carry USAF Blue

Scout rocket upper stages to launch small satellites. This plan was feasible but never carried out. If it had been, the X-15 would have become the first reusable launcher stage. It remains the only fully reusable vehicle system to have been to space and back.

Three X-15s were built, of which one was destroyed in a fatal accident and two are in museums. The X-15 project was abandoned in 1968. It had exceeded handsomely its original research objectives, and the priority at the time was the Apollo lunar programme. Some equipment developed on the X-15 was used in the early manned spacecraft, but the main beneficiary of the research was the Space Shuttle.

The X-15 was the first, and is still the only, sub-orbital spaceplane. (For present purposes, a 'spaceplane' is defined as a fully reusable vehicle capable of reaching space that lands like a conventional aeroplane, using wings for lift. This is equivalent to saying that a spaceplane is an aeroplane that can fly to and from space. 'Sub-orbital' means a trajectory that reaches space height, but not enough speed to stay in orbit like a satellite).

Inspired by the success of the X-15, most large aerospace companies in the early 1960s had design teams carrying out studies of reusable launch vehicles. A selection of European projects is shown in Figure 2.7, taken from [1]. These were called 'Aerospace Transporters'. My first job was on the Hawker Siddeley Aviation (HSA) project. Europe at the time lacked the organisation and political will to pursue such a project, and a vehicle on these lines has yet to be built.

The driving force behind these projects was Professor Eugen Sänger, who can be called the father of the spaceplane. His famous book on rocket flight engineering [2] was published in 1933. This was the first book on rocketry from an academic professional, and it includes a concept sketch of a rocket-powered aeroplane looking very like an X-15. He went on to design a sub-orbital bomber during World War II. This was way ahead of its time, but his technical report *A Rocket Drive for Long Range Bombers*, written with his wife Irene Bredt, was translated into English and Russian after the war and became highly influential [3].

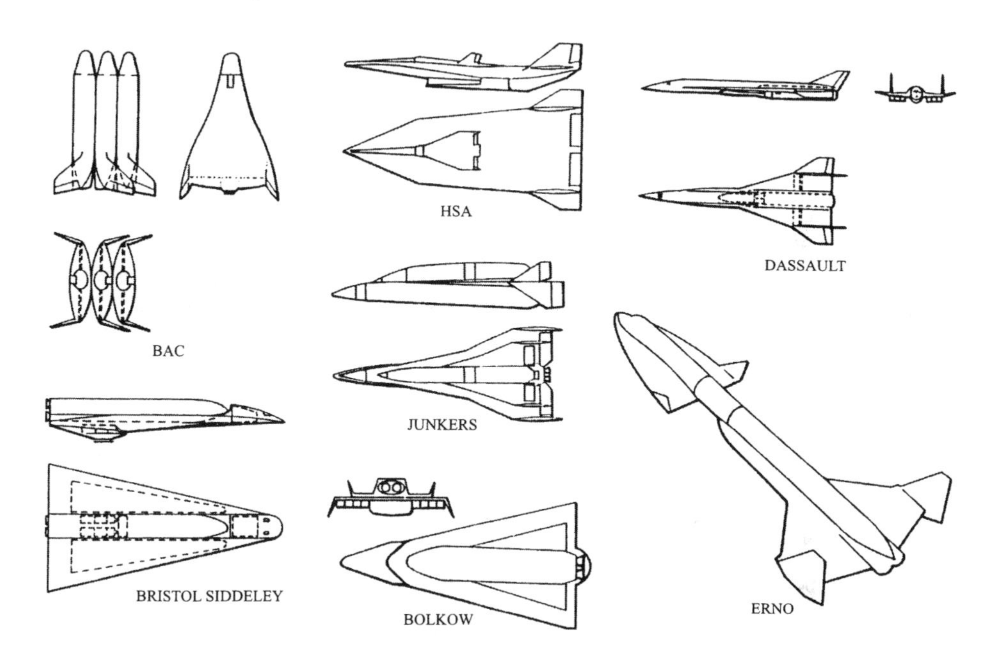

Figure 2.7 European Spaceplane Projects of the Mid-1960s
These projects were inspired by the success of the X-15, but have never been built.

There were several projects in the United States in the 1960s comparable to the European Aerospace Transporters but, again, none has been built. The consensus at the time was that an orbital spaceplane was the logical next step in space transportation, that it offered the prospect of greatly reduced launch costs, that it was feasible, but that it would need advanced technology and would be expensive to develop.

The United Kingdom perhaps came the closest to an operational sub-orbital spaceplane, more by accident than by design. The Saunders Roe SR.53 rocket fighter, Figure 2.8, first flew in 1957, two years before the X-15. The main engine was a rocket, but it also had a small jet to assist take-off and climb, and landing after the rocket fuel had been consumed. Designed to destroy Soviet nuclear-armed bombers, it had a rapid climb and a top speed of around Mach 2.

Figure 2.8 Saunders Roe SR.53 Rocket Fighter [GKN Aerospace]
The SR.53 rocket fighter first flew in 1957. It did not enter service, but could have been developed into a mature aeroplane with sub-orbital performance

The operational requirement changed, and the SR.53 did not go into production. It proved to be a practical aeroplane, with no particular handling problems in the air or on the ground. One of the two prototypes was destroyed in a fatal take-off accident, but this was not due to a fundamental design problem. When it was cancelled as a fighter, Saunders Roe proposed that it be converted for use as a high-speed research aeroplane [4]. The jet engine would have been deleted to make way for additonal rocket propellant, and it would have been air-launched from a Vickers Valiant bomber. With other modifications, it would have had a performance comparable to that of the X-15. This proposal attracted some government support, but not enough to make it happen.

The SR.53 was designed as a practical operational fighter. If it had gone into production, it would no doubt have matured to have reliability and operating costs like conventional jet fighters. With a modern higher performance rocket engine, it would have been capable of brief climbs to space height, which is a good indication of the feasibility today of a low-cost sub-orbital spaceplane.

Another relevant British project was the de Havilland Comet jet airliner fitted with de Havilland Sprite rockets to shorten the required take-off run at high air temperatures [5], shown in Figure 2.9. An experimental conversion made twenty flights in 1951 and 1952, including three flights in one day. At the time, no particularly difficult safety problems were envisaged or found with the rocket engines. To quote from the report, "The reaction to the rockets of the many passengers carried on the demonstrations was universally favourable. ...The passengers carried during the take-offs were of all types including chief designers of other aircraft companies, airline executives, test pilots, technicians and laymen."

Figure 2.9 de Havilland Comet with Sprite Rocket Motors to Boost Take-Off Performance. [The Philip Birtles Collection]
This aeroplane demonstrated, in 1951 and 1952, the potential of rocket motors to achieve safety and reliability adequate for airline use. Later versions of the Sprite rocket motors made less smoke.

The report does not say how many passengers were carried. Presumably, they were invited by the company to observe the flights from inside the cabin and give their comments. In a similar way, aircraft company employees are sometimes asked to take part in cabin mock-up trials of new airliners, to comment on comfort and convenience.

This Comet demonstration is probably the first and only time to date when passengers, albeit not fare-paying, have been carried in an aeroplane with rocket engines. While falling far short of spaceplane performance, it was a convincing indication of the potential of rocket engines to meet full airliner safety and maintenance standards.

Jet engines with higher thrust solved the take-off problem, and the Sprite Comet did not go into production.

Thus, the relevant technology for a useful small sub-orbital spaceplane to follow on from the X-15 had been demonstrated several years before the Shuttle was designed in the 1970s. Such a vehicle would have combined the performance of the X-15 with the ease of operation and safety that the SR.53 and rocket-assisted Comet would have achieved if they had gone into production and been developed to maturity. Contemporary studies left little doubt that vehicles of comparable practicability but capable of flying all the way to orbit were feasible and highly beneficial.

Why then was the Shuttle not made fully reusable? The original specification was indeed for a fully reusable vehicle, Figure 2.10, and it was relatively late in the programme when the decision to abandon full reusability was made.

The direct cause is simply explained. Cold War pressures were easing and the U.S. public had lost much of its early enthusiasm for space. President Nixon imposed a budget cut on NASA, who could no longer afford the development cost of the fully reusable Shuttle. The latter was in any case technically very ambitious, combining great size with great speed. NASA therefore sacrificed full reusability to contain development cost.

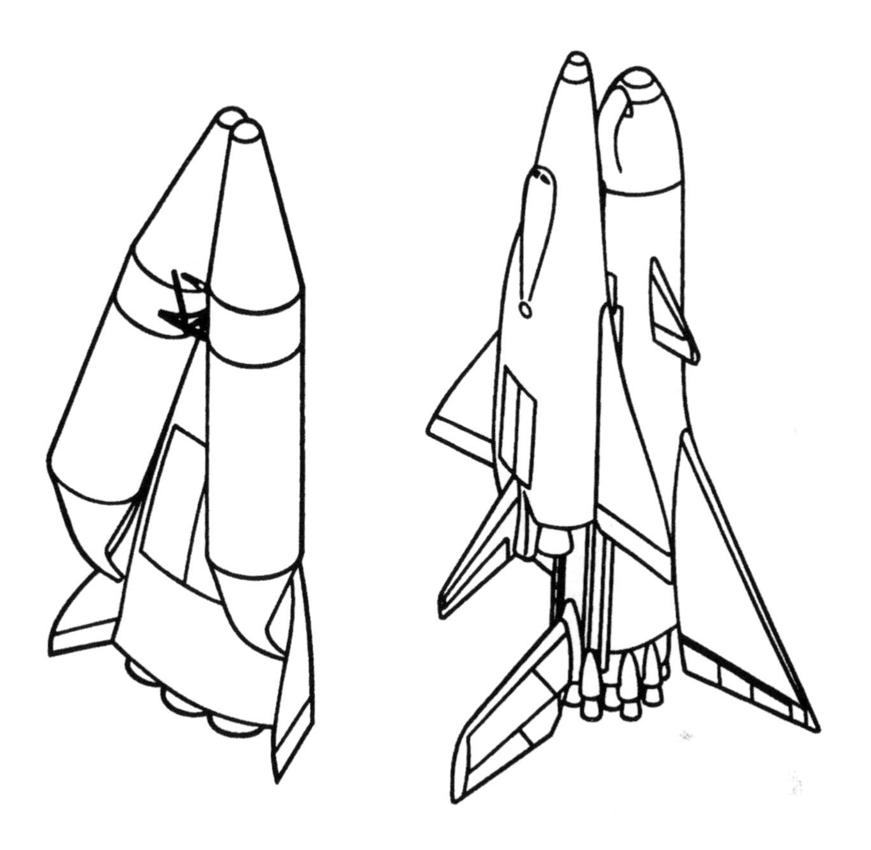

Figure 2.10 Lockheed Star Clipper and North American Design for Fully Reusable Shuttle [From Reference 6]

After a budget cut, NASA could not afford the original fully reusable design for the Space Shuttle (right). The Star Clipper (left) would have been far less expensive to develop and to operate than the design that was built.

Less obvious is why NASA did not maintain full reusability, and reduce the payload size to contain cost. The Lockheed Star Clipper, Figure 2.10, designed by Max Hunter, was a leading contender for the latter solution (although its propellant tanks were expendable). There was indeed a small lobby in favour of the smaller shuttle, and an independent report commissioned by the U.S. Office of Management and Budget and carried out by Mathematica Inc recommended such a project.

A good history of this period [6] quotes the director of the Mathematica studies, Klaus Heiss, as saying in the context of the smaller project: "For a long time some people over there [at NASA] kept seriously telling us 'We can't go that route, because we've got to do something for the Marshall Space Center as well as something for the Houston space center.' " This is a clear indication of how space policy at the time was driven by politics rather than by engineering and commercial common sense. The case for the small shuttle was swamped by the politics of mega-projects.

A small shuttle like the Star Clipper could have been built at a fraction of the cost of the eventual design. It would have been a straightforward extension of X-15 technology. With such a development, together with the mighty Saturn launcher and the Skylab space station, NASA would by the late 1970s have had the three basic types of vehicle needed for a complete orbital infrastructure, i.e., spaceplane, heavy lift vehicle, and space station.

The Saturn heavy lift vehicle would have launched Skylab space stations, and the much smaller Star Clippers would have been used for regular supply flights to these space stations. The Star Clipper would have led naturally to a more economical design, and the Saturn would have been developed progressively into a reusable heavy lift vehicle. The manufacturers of the three stages of Saturn (Boeing, North American, and Douglas) all had proposals during the 1960s for reusable derivatives. These proposals included adding a parachute recovery system to the lower stage, and then to the upper stages, followed by wings on the lower stage and so on.

Costs would have reduced rapidly due to the virtuous circle of reusability, maturing technology and increasing utilisation. At some stage in this process, an entrepreneur would probably have realised that mature developments would lead to costs low enough for profitable space tourism, and we could have had a flourishing space tourism industry a decade or more ago. This is the 'X-15 Way' to the 'Spaceplane Space Age' shown later in Figure 4.2.

But NASA was not geared up for incremental advances. They had become culturally and politically attuned to deca-billion-dollar projects.

Thus it came about that the Space Shuttle, intended to greatly reduce the cost of spaceflight, became just as expensive and risky as the manned spacecraft on expendable launchers that it replaced, and that an opportunity to bring in the spaceplane space age by the 1980s was missed because of a preoccupation with the politics of maintaining a high budget.

The Space Shuttle has indeed done great things, but even greater things could have been achieved at far less cost.

CHAPTER 3 - RECENT DEVELOPMENTS

Since the first flight of the Space Shuttle in 1981, there have been numerous proposals for reusable launch vehicles. Only one of these has progressed as far as a flying demonstrator. The McDonnell Douglas DC-X was an experimental vehicle to test some of the technology for the proposed Delta Clipper single-stage reusable launcher. It took off and landed vertically using rocket engines, the first vehicle to do so. It made twelve flights from 1993 to 1996. It made two of these flights within a 26-hour period, to demonstrate rapid turn-around with a small crew. On the last landing, a landing gear strut failed to extend, the vehicle tipped over, the liquid oxygen tank exploded, and the vehicle was destroyed. This project cost about $120 million. The DC-X and some other recent reusable launch vehicle projects are shown in Figure 3.1.

Two more projects came close to flying. The X-33 was an experimental sub-orbital spaceplane designed to test some of the technology of the proposed Lockheed Venture Star single-stage reusable launcher. It was designed to be unpiloted, to have vertical take-off and horizontal landing, and to reach a maximum speed of Mach 15, which is well over one half of orbital velocity. Largely funded by NASA, this $1.5 billion project was cancelled when under construction in 2001 because of growing engineering problems, cost overruns, and the safety issues of flying large unpiloted vehicles over the continental United States.

The Orbital Sciences X-34 was an unpiloted experimental sub-orbital spaceplane designed to test technology for reusable launch vehicles. It was less ambitious than the X-33 and would have been somewhat larger, faster, and higher than the X-15. It would have been launched from a converted Lockheed L-1011 airliner and had a development cost of around $250 million. It was cancelled by NASA in March 2001 when almost complete but before it could fly. As with the X-33, one of the reasons for the cancellation was the safety issue of large unpiloted aeroplanes over the United States.

The DC-X demonstrator for a vertical take-off, vertical landing reusable launcher made 12 flights from 1993 to 1996. It cost about $120 million.

The X-33 was a demonstrator for a single stage to orbit reusable launcher. It was cancelled by NASA in 2001.

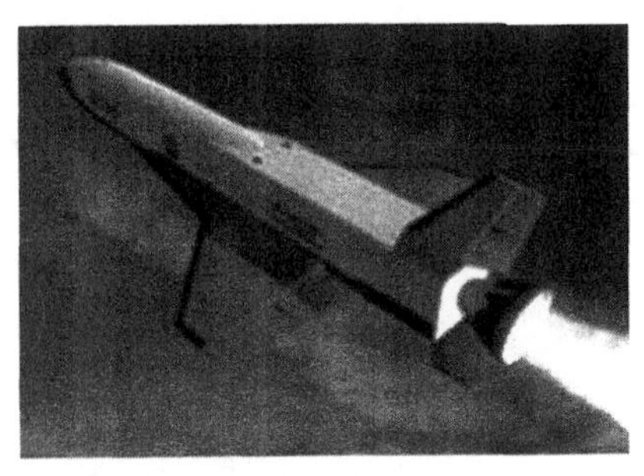

The X-34 was a less ambitious demonstrator for reusable launch vehicle technologies. It would have cost some $250 million, but was cancelled by NASA in 2001, when almost complete.

The Kistler K-1 is a $1.5 billion private venture reusable launch vehicle, in search of funding.

Figure 3.1 Some Recent Reusable Launch Vehicle Projects [NASA and Kistler]
Since the Space Shuttle first flew in 1981, these are the reusable launch vehicle demonstrators or prototypes that have flown or come closest to flying.

Since the 1960s, some work on spaceplanes has been carried out in Europe and elsewhere, but less than in the United States. The European project on which most work has been done is the Sänger spaceplane, described later.

In parallel with these official projects, there is a growing band of small start-up companies trying to raise the funding for low-cost reusable launch vehicle development. One of these projects has progressed beyond the drawing board. The Kistler K-1, Figure 3.1, is a large privately-funded reusable launch vehicle. Recovery is by parachute, with air bags to prevent damage on landing. At the time of writing, Kistler have obtained about two thirds of the required $1.5 billion development funding. If development plans succeed, the K-1 will be the first reusable launch vehicle, although it will not be a true spaceplane because of the parachute recovery. Nevertheless, it offers the prospect of reducing launch costs by up to five times (my estimate), once well down its various learning curves.

Most of these small start-up companies see space tourism as the big market for their spaceplanes, and there is now a small but active and growing space tourism movement. Professional journals over the years have published several papers, e.g., [8, 9, 10, and 11], showing that, if spaceplanes were to achieve airliner maturity, the cost of space travel would be reduced a thousandfold, which many ordinary people could afford. Three of these papers are from the 1950s and 1960s, when the general assumption was that the large new market for spaceplanes would be high-speed air transport, but space tourism is now generally seen as an earlier business opportunity. At the time they were written, these papers were taken as an indication of long-term potential rather than as the basis for immediate action. They still, however, provide a theoretical background to present claims about the low-cost potential of spaceplanes.

These start-up companies claim that a spaceplane could be built at far less cost than NASA is assuming in its plans. NASA recently started a $4.5 billion programme called the Space Launch Initiative, which is a five-year programme to prepare the technology for a reusable launch vehicle. The $1.5 billion needed for the Kistler K-1 project, for example, is one third of the cost of this programme, which will not even lead to an operational vehicle. At least one of these claims for low cost has received government endorsement. In 1993, Bristol Spaceplanes Limited received a contract from the European Space Agency for a feasibility study of the Spacecab spaceplane [33]. The conclusion was that such a vehicle could be built with existing engines and proven materials and systems for a development cost of $2 billion. The UK Minister of Space asked for an independent review of this work, which broadly agreed with the conclusions [12]. Thus, a study paid for by the European Space Agency and broadly endorsed by the UK government supports the low-cost case.

Most start-up companies would probably say that $2 billion is on the high side for orbital spaceplane development, and they may well be correct. One promising project is the PanAero Inc X Van [32]. The cost of three X Van operational prototypes is quoted as $130 million. The X-33 would have cost about $1.5 billion if it had not been cancelled, and it was a more ambitious design than a two-stage spaceplane like Spacecab or X Van.

Most of these companies are competing for the X-Prize, which is a $10 million prize for the first non-government organisation to demonstrate a reusable and piloted sub-orbital vehicle capable of carrying three people to 100 km. To win the prize, the vehicle must make two 100 km flights within two weeks [13].

The idea behind the prize is to stimulate spaceflight in the same way that the Orteig prize stimulated commercial flying. Charles Lindbergh's non-stop flight from New York to Paris in 1927 won the prize and transformed United States air transport and private flying. To quote from an authoritative history: "The United States was already well advanced in service and racing aircraft, and in the operation of mail routes; but it took the romantic impact of Lindbergh - similar in its tremendous impact [in the United Kingdom] to Blériot's Channel crossing of 1909 - to make the nation truly air-minded, and so create the financial and technical climate necessary for the large-scale development of aviation" [14].

There is a major difference between the Orteig prize and the X-Prize. Lindbergh's achievement was at the leading edge of technology whereas the X-15 was achieving X-Prize performance 40 years ago. (It only had one seat, but could have been modified to carry two more in its pressurised equipment bay, to meet X-Prize conditions.)

The key phrase in the X-Prize rules is "non-government". The aim of the prize is to demonstrate the prospect for low-cost spaceplanes and space tourism, and thereby either induce a change of government policy or enable the full development funding to be found by the private sector.

If the private-sector start-up companies are correct, a spaceplane could be built for less than half the cost of the NASA programme just to prepare the way for the start of development.

CHAPTER 4 - WAY AHEAD

This chapter summarises a predicted way ahead for space transportation, based on the more promising designs from the 1960s and later ideas from the private sector. It minimises development cost and risk by using technology developed since the 1960s for other projects. It involves a step-by-step development sequence of progressively more advanced vehicles, each one paving the way for the next in terms of technology, market build-up, and financial credibility. This way ahead is called 'The Aeroplane Approach' because the vehicles are as much like conventional aeroplanes as practicable. Clearly, other ways ahead are possible, and later chapters will provide evidence to support the prediction that the private sector approach will prevail, and that a way ahead along the general lines of that that described here is likely to happen soon. The development sequence is shown in Figure 4.1.

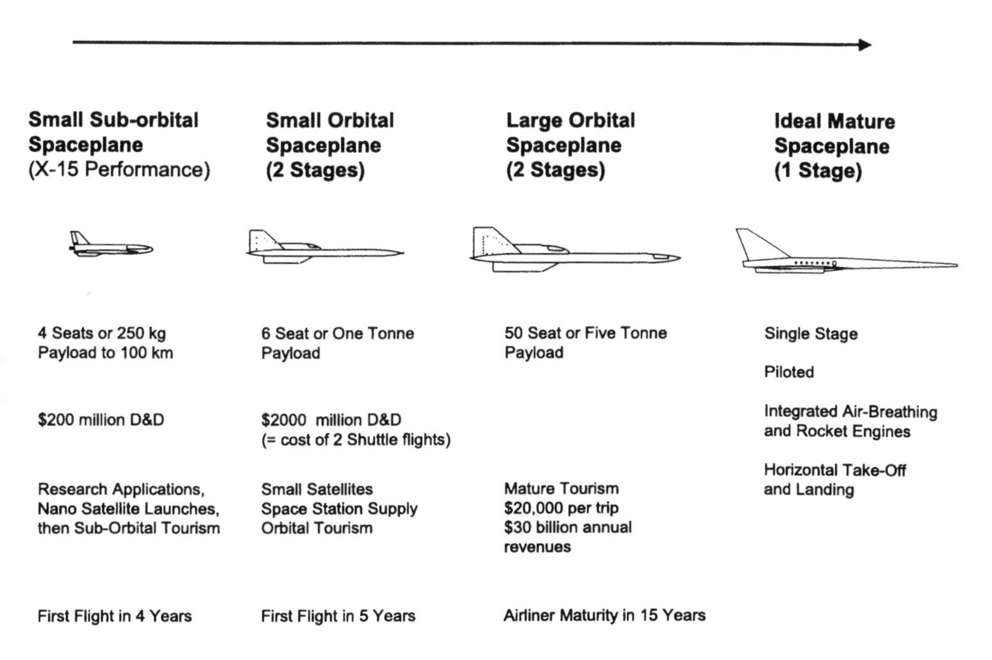

Figure 4.1 Spaceplane Development Strategy (The Aeroplane Approach)
A way ahead for spaceplane development along these lines is likely to start soon, leading to airline operations to orbit within fifteen years.

The first vehicle is a small sub-orbital spaceplane with a performance comparable to that of the X-15. It is capable of several brief zoom-climbs to space height per day, landing back at or close to the airfield that it took off from. It combines the performance of the X-15 with the ease of operation and safety that the SR.53 and Sprite Comet would have achieved if they had gone into operational service. Its basic technology is forty years old.

A prototype is built in four years for a cost of about $200 million using existing engines and proven technology. Early flights carry test instrumentation, as well as scientific experiments to defray the cost of the test flying. After sufficient flights to demonstrate safety and reliability, it is certified for carrying passengers on space experience flights. Passengers feel weightless for about two minutes. They see the curvature of the Earth clearly, and an area several hundred kilometres across at one time. They see the sky go dark with bright stars even in daytime. They know that they have been to space. The fare on early passenger flights is around $100,000. The market at this cost is about 10,000 people per year, resulting in an annual demand of about $1 billion. With economies of scale and with maturing technology, the fare comes down to a few thousand U.S. dollars. In about ten years, most people who are reasonably fit and well-off will be able to take brief trips to space.

This 'reinvented X-15' conclusively demonstrates the advantages of aeroplane-like space vehicles, starts to build up the market for new commercial uses of space, and provides a focus for maturing the technology. It thereby paves the way for the remaining vehicles on the development sequence.

The second vehicle is a small orbital spaceplane, using the best features from the 1960s designs and using technology developed since then for other projects. It carries a satellite in the one tonne class, or six people. These are space station crew, mechanics to service a satellite in orbit, or passengers on pioneering orbital tourism flights. A prototype is built in about five years using existing engines and proven technology for about $2 billion, which is roughly the cost of two Space Shuttle flights. Government space agencies recover its development cost in about one year, by using it instead of the Shuttle for supply flights to the International Space Station.

The third vehicle is an enlarged development of the small orbital spaceplane, capable of carrying a five-tonne satellite or 50 people. The largest market for this vehicle is carrying passengers to space hotels. The cost of a stay of a few days when the system is fully mature is around $20,000. Costs that low require airline-like operations, with spaceplanes capable of several flights per day to space and having a long life and low maintenance cost. This standard of maturity is reached in about fifteen years. The pacing item is the development of rocket engines with long lives and low maintenance costs, comparable to airliner jet engines.

Space hotels are equipped with viewing rooms providing superb views of Earth, planets, stars, and galaxies. They have large gyms for low-g flying and swimming. More than one million people per year are prepared to pay $20,000 for a once-in-a-lifetime visit to space, which requires a fleet of more than 50 spaceplanes. Annual revenues from space tourism are more than $20 billion.

The vehicles on the development sequence so far have been designed for low development cost and risk, which implies the use of existing technology. The penalty is the need for the small and large orbital spaceplanes to have two stages - a carrier aeroplane and an orbiter. This increases complexity and operating cost. The final vehicle on the development sequence is called the 'Ideal Mature Spaceplane', and uses technology now being researched to allow the use of a single stage.

The potential revenues from tourism should provide most of the funding for this development sequence. However, the taxpayer will probably have to fund the early stages until the private sector has the confidence to take over the funding. A prototype sub-orbital spaceplane making regular flights to space should provide this confidence. The business potential for spaceplanes should then become straightforward to explain. Thus, the cost to the taxpayer of the Aeroplane Approach should be no more than the $200 million to fund the prototype of the small sub-orbital spaceplane. At the very most, the cost to the taxpayer will not exceed the $2 billion for the small orbital spaceplane. By contrast, NASA thinks in terms of a mature spaceplane in about forty years [15]. The present NASA annual budget for manned spaceflight is approximately $5 billion. In the absence of strong external pressure, this would probably carry on for the next forty years and the total spend in that period would be well over $100 billion.

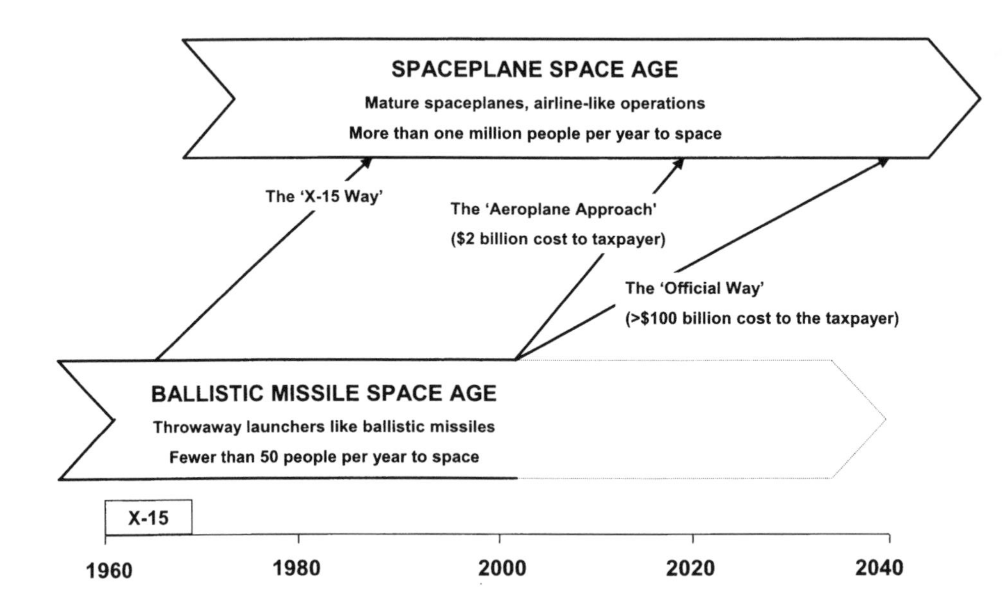

Figure 4.2. A Transformation of Space Operations is Likely Soon
The 'Aeroplane Approach', based on updated ideas from the 1960s, leads sooner and at lower cost to the Spaceplane Space Age than the 'Official Way'.

Figure 4.2 compares three ways ahead for space transportation. The 'X-15 Way' is what could have happened if some of the more promising projects in the 1960s had been developed then, leading to a mature spaceplane by about the mid-1980s. The 'Aeroplane Approach' is that described in this chapter, which should lead to a mature spaceplane in about 15 years for a cost to the taxpayer not exceeding $2 billion. The 'Official Way' assumes extrapolation of current trends in space policy, leading to a mature spaceplane in about 40 years at a cost to the taxpayer of more than $100 billion.

As well as slashing the cost of spaceplane development, the private sector could greatly reduce the cost of space stations. The International Space Station, Figure 4.3, is now being assembled in orbit, following a long gestation period. Led by NASA, it will have modules from Brazil, Canada, Europe, Japan, Russia, and the United States. It is difficult to find an estimate of the total cost of the International Space Station, but it is probably around $100 billion [7].

Figure 4.3 International Space Station: Artist's Impression of System When Complete [NASA]
The International Space Station is being assembled in orbit.
The total cost is likely to be around $100 billion.

The NASA annual budget of around $2 billion for its contribution to the International Space Station would purchase more than 20 Russian Soyuz mini space stations per year. Many commentators have pointed out the irony that Russia is far more entrepreneurial with its space assets than the United States For example, Russia encouraged Dennis Tito to become the first space tourist against strong opposition from NASA.

It is all but obvious that better science could be obtained for about 10% of the cost of International Space Station by having four or more smaller space stations using modules of International Space Station and Soyuz design. There are four main space science disciplines; astronomy, atmospheric science, Earth science, and microgravity, each requiring a variety of scientific instruments. Each instrument has an optimum orbit. Some should be high, others low; some circular, others elliptical; some should be in equatorial orbit, others in polar orbit, and so on. Four smaller space stations in different orbits offer closer to optimum orbits for the instruments. One large space station involves more detrimental compromises.

Useful smaller space stations each dedicated to a particular discipline could be built and launched for less than $1 billion each (equivalent in cost to more than ten Soyuzes). With a small spaceplane to provide subsequent supply operations, the annual cost of each would be less than $100 million. The total cost of four such space stations over ten years would then be less than $8 billion. Including the cost of developing the spaceplane, the cost would be less than $10 billion, some 10% of International Space Station cost.

These are preliminary estimates. A good way of obtaining a more reliable indication of cost would be competitive bidding by the private sector to provide specified accommodation in orbit for scientists and instruments for a given number of years. Such a competition would require a major change in government policy.

The 'Aeroplane Approach' as described here is clearly far more beneficial than the 'Official Way'. The rest of this book describes the analysis behind the prediction that the ‘Aeroplane Approach’ will start soon, leading to airline-like flights to orbit in about 15 years.

PART 2

Spaceplane Potential

CHAPTER 5 - SPACEPLANE LOW-COST POTENTIAL

A key to the Aeroplane Approach to space transportation is the potential of mature developments of spaceplanes on the drawing board today to reduce the cost of sending people to space by a factor of one thousand.

A good starting point is to consider the present cost of sending people to space. Each flight of the Space Shuttle costs between $500 million and $1.5 billion, depending on what is included [16, 18]. For convenience, we will use a figure of $1 billion, which is some 10,000 times more than the cost of a long-distance flight of a Boeing 747.

This high cost is due to expendability. The Orbiter is reusable, but the large External Tank burns up on re-entry, and the two Solid Rocket Boosters are recovered at sea and re-cycled, with little cost saving compared with new ones. The Shuttle is not a spaceplane according to our definition because it is not fully reusable. The only other semi-reusable launcher in service - the Orbital Sciences Pegasus - uses a converted airliner (a Lockheed L-1011) as a mobile launch platform. The other launchers are all totally expendable. Ariane 4, shown in Figure 5.1, is typical of these.

Figure 5.1 Ariane 4 [Arianespace]

Ariane is a successful launch vehicle in present service.
It can fly once only. Its technology derives directly from ballistic missiles.

The technology of all these expendable stages derives directly from ballistic missiles.

The Shuttle usually carries seven people, and the cost per person per flight is therefore around $140 million. The Shuttle could, if required, be configured to carry about 50 people in place of cargo, so the cost per seat could come down to around $20 million.

Figure 5.2 The Russian Soyuz Launcher and Manned Spacecraft [Commercial Space Technologies]
The Soyuz manned spacecraft is available, including launch, at a cost in the region of $60-100 million.

The other people-carrier, also expendable, is the Russian Soyuz, Figure 5.2, which can carry three people. It is not easy to find the cost of Soyuz because the figures are not published and because of the variable currency conversion rate. Starsem market the Soyuz launcher in the West, but not the manned spacecraft, and quote a cost for a launch of $30-50 million. If we use a convenient rule of thumb and assume the same cost for the manned spacecraft, the total becomes $60-100 million. From unofficial sources, the lower figure is probably more accurate.

Thus, in very round numbers, the present cost per person for a few days in space is $20 million, and this indeed is the sum that the first true space tourist, Dennis Tito, paid for his visit to the International Space Station in the year 2001.

One might expect that the prospect of a thousandfold reduction in cost using mature developments of existing technology would be receiving priority attention from space agencies and aerospace corporations. That this is not the case is the main reason for this book.

Perhaps the most relevant recent cost estimate by a large corporation is that of Messerschmitt-Bölkow-Blohm (MBB, now part of EADS) for their Sänger spaceplane, shown in Figure 5.3. Sänger is one of the most thoroughly studied spaceplane projects, and has many of the key design features needed for commercial success. For these reasons, Sänger is referred to several times in this and later chapters.

The design of Sänger dates back to the early 1960s when Junkers, a predecessor company of MBB, was involved in the European Aerospace Transporter studies. Work on Sänger peaked in the late 1980s and early 1990s when it was the centrepiece of the German hypersonic research programme, but it is no longer being actively pursued.

Sänger is a two-stage vehicle, consisting of a Carrier Aeroplane and an Orbiter (my nomenclature). The Carrier Aeroplane takes off from a runway, flies to the required orbit plane, and accelerates to Mach 7, which is just under one third of orbital velocity. The Orbiter then separates and accelerates to orbit. The Carrier Aeroplane flies back to the airfield that it took off from, and is prepared for the next launch. The Orbiter delivers its payload to orbit, reenters the atmosphere, and flies back to base.

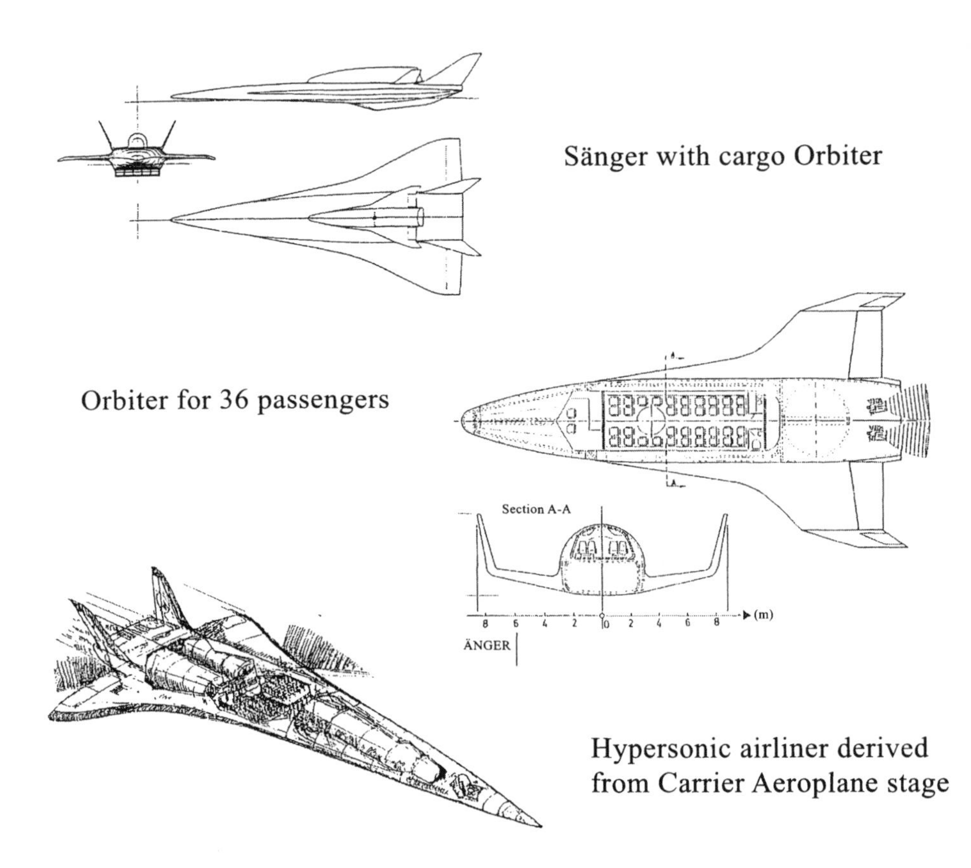

Figure 5.3 The MBB Sänger Spaceplane [EADS]
Sänger was thoroughly studied in Germany in the late 1980s and early 1990s. It had many of the key features for commercial success.

There are two versions of the Orbiter, one for passengers and one for cargo. Figure 5.3 shows the version designed for carrying 36 passengers to and from orbit.

Sänger is long and slim, like Concorde, to reduce drag at high speed. It has a take-off mass comparable to that of a Boeing 747. Two major differences are that Sänger has two stages - the Carrier Aeroplane and Orbiter - and uses liquid hydrogen fuel. These features are to reduce the required propellant mass fraction to an achievable level. For a single-stage vehicle to reach orbit using rocket engines with kerosene fuel and liquid oxygen propellants, the propellant mass would need to be 94% of the take-off mass, which is far too high for a practical spaceplane.

Hydrogen fuel has more energy per unit mass than kerosene, and having two stages decreases the fuel fraction required for each stage.

Even with these palliatives, Sänger has to carry more fuel than the 747 and can therefore carry about twelve times fewer people - 36 compared with around 420. This clearly increases the cost per seat.

MBB also designed a supersonic airliner based on the Carrier Aeroplane with a cruising speed of Mach 4.4, which is just over twice the speed of Concorde. Other manufacturers have explored such commonality between high-speed airliner and launcher carrier aeroplane as a means of spreading development cost, and Sänger is the most recent example.

The estimated cost per flight of the airliner version of Sänger was $125,000 in 1986 [17, 18], which would be around $200,000 today. This is somewhat less than twice the cost per flight of a Boeing 747, and is broadly in line with estimates over the years for a hypersonic airliner with mature technology. Assuming equal technical maturity, there is no obvious reason for the Carrier Aeroplane to cost much more per flight. It is the same size and has a lot of design commonality. It flies faster but for a shorter time, which should approximately balance out. The Orbiter is smaller but more advanced, and if for the sake of a quick-look estimate we assume that these factors also cancel out, the cost per flight of the Orbiter would also be around $200,000, to give a total of $400,000. With 36 seats, the cost per seat to orbit is then just over $10,000, or 2000 times less than the cost today. This conclusion is in line with those of the technical papers mentioned earlier [8,9,10 and 11]. The underlying assumption is that the launcher technical maturity is comparable to that of an airliner. This is an important assumption, the implications of which will be considered later.

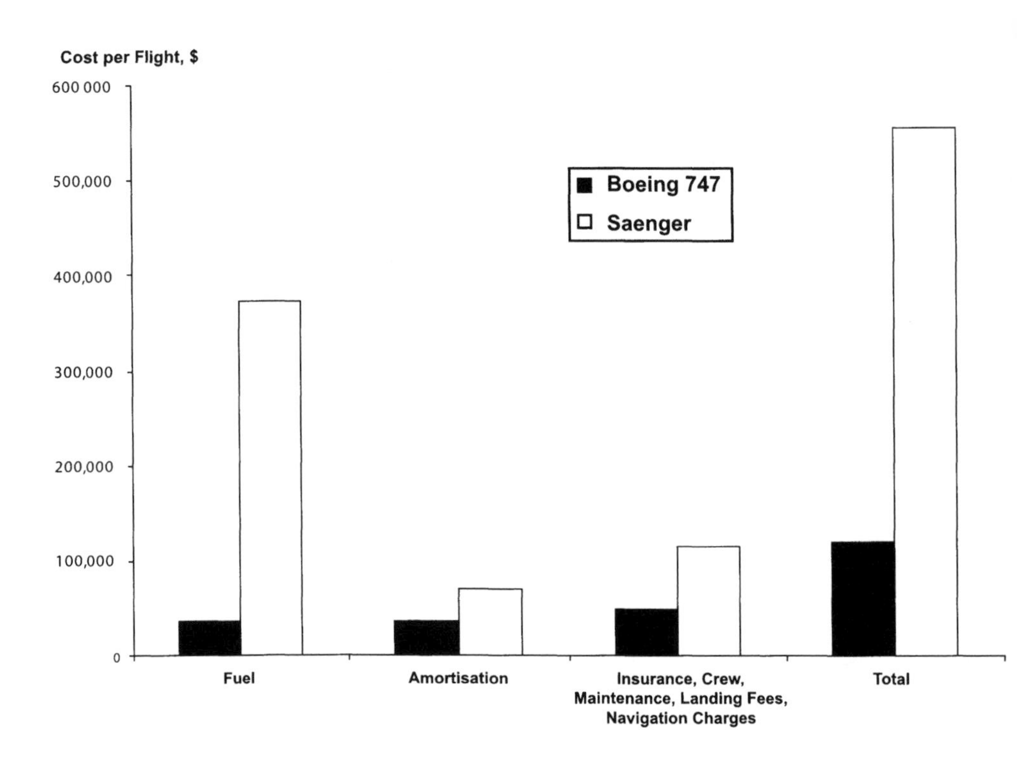

Figure 5.4 Cost Comparison Between Mature Sänger and Boeing 747
If developed to airliner maturity, the cost per flight of Sänger would be some $550,000 and the cost per person to orbit around $15,000.

This is clearly a very rough estimate, but it does indicate the low-cost potential of a mature spaceplane. A more detailed estimate is given in Appendix 1 and summarised in Figure 5.4. The basis is to estimate mature Sänger costs by scaling those of a Boeing 747. The intention is to illustrate the potential cost of spaceplane flight, assuming mature technology. The costs therefore assume that Sänger has the same life and turn-around time as the 747.

With these assumptions, Sänger has a cost per flight about five times greater than a 747. It carries about 12 times fewer people, and the cost per seat is therefore some 60 times greater, at around $15,000.

Nearly 60% of the total cost is due to the hydrogen fuel for the Carrier Aeroplane. Hydrogen is some ten times more expensive per unit weight than kerosene. The other cost items could therefore change significantly without greatly affecting the total.

Sänger was not optimised for a high launch rate. If it had been, it would probably have used kerosene fuel for at least part of the ascent, in order to reduce fuel cost. A preliminary cost estimate of a project that was so optimised, Spacebus, indicates a cost per seat to orbit of around $5000 [19]. This large reduction compared with Sänger is mainly thanks to using kerosene fuel for a large part of the Carrier Aeroplane ascent. It is also thanks to carrying more passengers (50 compared with 36), made possible by different design features. Even so, the cost per flight in Spacebus is more than twice that of a 747, and it carries eight times fewer passengers.

The estimate for Sänger probably represents an upper limit for the cost per seat to orbit of a spaceplane using mature developments of technology more or less available today, and Spacebus a lower limit. It is difficult to be more precise at this stage. However, when a thousand times reduction in cost can be reliably predicted, a factor of two or three either way at this stage is relatively unimportant.

Thus, in round numbers and based on a robust estimate, a thousandfold reduction in the cost of sending people to space can be achieved with a mature spaceplane. The main requirement for this to happen is for the technical maturity to approach airliner standards.

The winning equation is therefore:

Reusability + Maturity = 1000 times launch cost reduction

Sänger uses new turboramjet engines to Mach 7, whereas the fastest yet achieved by an air-breathing engine is the Mach 4.3 of the Lockheed X-7 test missile of the 1950s. The use of such advanced engines partly explains the ten-year development timescale quoted by MBB.

The quoted development cost was around $17 billion, which is two or three times that of a large new airliner. The projected market for the launcher was twelve flights per year for satellites and supplying space stations. This compares with more than one million flights per year for a successful airliner design.

MBB were unable to attract the required funding, probably because of this combination of risky technology, high development cost, and small market. We clearly need to greatly reduce development cost and/or to find large new markets.

There has been so much design work and flight demonstration on spaceplanes that there is no sensible doubt that the prototype of a vehicle on the general lines of Sänger could be built if required. A smaller and less advanced design with lower development cost might be more appropriate for the first spaceplane, but Sänger serves as a well-established datum point.

There is equally little doubt that Sänger would achieve costs comparable to those estimated here if developed to airliner maturity. Such development would be expensive, so the main requirement for it to happen is a large enough commercial demand to provide the operating experience and funding for continuous product improvement towards airliner standards of safety, life, maintenance cost and turnaround time.

To take full advantage of low-cost spaceplanes, additional vehicles are required, as discussed in the next chapter.

CHAPTER 6 - ORBITAL INFRASTRUCTURE

To provide a complete low-cost service to orbit, other types of vehicle are needed to supplement spaceplanes. Space stations will be needed to serve as observatories, factories, hotels, fuel depots, and assembly depots for large interplanetary craft. These space stations will be assembled from modules launched separately from Earth. Large launchers, usually called heavy lift vehicles, will be needed to launch these modules and other payloads too large for the spaceplanes. The main use of spaceplanes will then be supply flights to the space stations, ferrying replacement crews, scientists, passengers, workers, spare parts, and consumables.

This combination of spaceplane, space station, and heavy lift vehicle has been called a 'mature orbital infrastructure' and is shown in Figure 6.1. A useful analogy is a large offshore oil platform that needs a large ocean-going tug to position it just once, and then relies on helicopters or smaller boats for supply operations.

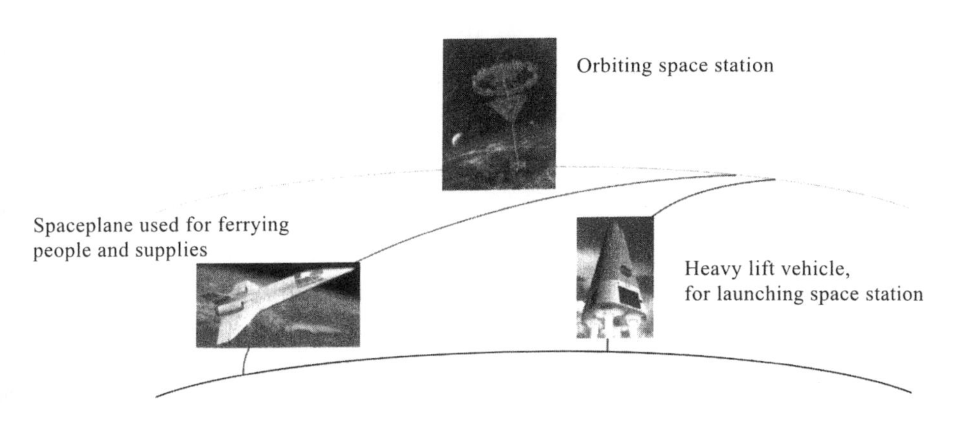

Figure 6.1 Mature Orbital Infrastructure
Spaceplanes are the enabling development for a mature orbital infrastructure.

This idea of a complete orbital infrastructure is by no means new. The rocket pioneer von Braun and others carried out realistic studies in the early 1950s [20]. NASA came close to achieving it thirty years ago with the Skylab space station, Saturn heavy lift vehicle and the X-15 sub-orbital spaceplane - but did not close the gap by developing an orbital successor to the X-15, as discussed earlier.

Present space stations are very expensive to build and operate. As mentioned earlier, the International Space Station will have a total cost probably approaching $100 billion. The Russian Soyuz manned spacecraft is more representative of a production standard because it has been built in greater numbers. It can be adapted for used as a small space station, carrying a crew of three. It weighs seven tonnes and costs approximately $30 million, depending on currency conversion rates as mentioned earlier. The cost per tonne is therefore some $4 million, which is very roughly five times that of an airliner.

Neither the International Space Station nor Soyuz represents the mature space stations that will be possible when low-cost access by spaceplane becomes available. In estimating the manufacturing cost of space stations when fully mature, the 'engineering standard' has to be assessed. By this is meant the lightness or the sophistication of the design, as indicated by the cost per unit weight. With different types of building or vehicle, this tends to increase with top speed, or with installed power, or with the cost penalty for excess weight.

Present human accommodation can be ranked in ascending order of engineering standard as land static (e.g. hotels), sea static (floating buildings), sea mobile (cruise liners), land mobile (coaches and trains), air (airliners) and space (space stations). The cost per unit weight of an airliner is far greater than that of a hotel, for example. The question is where space stations will fit in this spectrum when fully mature.

Present space stations require a very high engineering standard because:

- They have a high political profile

- The production rate is very low - of the order of one per ten years in the United States
- The technology is not mature
- Each tonne of extra weight costs more than $10 million to launch
- The call-out charge for a plumber or electrician to fix a fault is around $1000 million, the cost of a Shuttle flight.

Every predictable eventuality therefore has to be designed for, and the crew trained for, within very tight mass budgets.

When mature spaceplanes provide low-cost access on demand, none of these factors will apply, and space stations will soon take on their 'natural' engineering standard.

A space station is essentially simpler than an airliner of comparable size. It does not require large engines, wings, tail, flying controls, flight instruments or landing gear. It consists in essence of an airliner fuselage with more sophisticated systems for environmental control and life support; and fitted with an attitude and orbit control system, and solar panels.

A space station is inherently safer than an airliner. One of life's few certain predictions is that space stations will not hit the ground in an unplanned fashion at short notice. They are astronomical objects of known orbit. They will not get lost. They are not affected by the weather. They can be repaired in flight. There will be time to fix mechanical failures. In a grave emergency, a space station can be evacuated. Most of the factors that cause fatal airliner accidents will therefore not give rise to urgent problems on space stations. There will, however, be some new hazards resulting from operating in space but, as discussed later, it should be straightforward to manage these risks to provide acceptable safety.

An airliner makes typically 30,000 flights in its economic life. Each kilogram of extra structure weight means 1 kg less payload on many flights, which adds up to a considerable loss of revenue over the life of an airliner fleet. In contrast, a space station has to be launched only once. Thus, when low-cost launches become available, the cost penalty for an extra kilogram of weight will be far less for a space station than for an airliner.

For all these reasons, the engineering standard of mature space stations is likely to be somewhat lower than that of airliners. However, to be conservative, the cost estimate later in this chapter will assume a somewhat higher cost per tonne.

Nuclear submarines require standards of safety and life support comparable with those of space stations, and these cost typically six times less per tonne than airliners. They would cost even less if produced in comparable numbers. Space hotels are not going to be built as heavily as submarines, but the analogy illustrates the point that the cost of providing safe accommodation in a hostile environment depends largely on the cost of getting there. Mature spaceplanes will therefore slash the overall cost of space stations.

To gain insight into the relative importance of the cost of spaceplanes, space stations, and heavy lift vehicles, Table 6.1 shows a first order estimate of the cost per person for a few days in a space hotel when all systems are fully mature.

On this basis, the annual cost per tonne of space hotel is $242,000, and that per tourist place is $363,000, assuming the hotel weighs 1.5 tonnes per place. Assuming that the average visit is three days and making a small allowance for cleaning and maintenance between visits, 100 visitors per year use each place, and the direct cost of the hotel for a short visit is $3630. Adding the cost of a ferry flight in a mature spaceplane and including an allowance for the transport of supplies, the total cost of a few days in a space hotel is then around $11,000.

Table 6.1. Cost of Three-Day Visit to Mature Space Hotel

ITEM	VALUE	NOTE
ASSUMED BASIC DATA		
Hotel Weight per Tourist Place	1.5 tonnes	1
Tourist per Place per Year	100	2
Number of Tourists per Crew Member	2	
Hotel Production Cost per Tonne	$1,000,000	3
Hotel Launch Cost per Tonne	$50,000	4
Hotel Capital Cost per Tonne	$1,050,000	
ANNUAL HOTEL COSTS PER TONNE OF HOTEL		
Amortisation	$105,000	5
Insurance	20,000	6
Maintenance	50,000	7
Crew	67,000	8
Total Annual Hotel Cost per Tonne	242,000	
COSTS PER TOURIST VISIT		
Hotel	3,630	9
Ferry Flight	5,000	10
Supplies Transportation Cost	2,500	11
Total Cost per Visit	$11,130	

NOTES

1 This is the weight of a medium saloon car or ten times the weight per passenger of a typical airliner fuselage.

2 Two occupancies per week (Average stay of three days plus one day for cleaning, etc.)

3 Forty percent more than that of a typical airliner (to give a conservative round number)

4 Spacebus launch cost per tonne of payload

5 10% of capital cost

6 2% of production cost

7 5% of production cost

8 $200,000 per year per crew member, including overheads, ÷ 2 tourists per crewmember ÷ 1.5 tonnes of hotel per tourist

9 $242,000 per tonne per year x 1.5 tonnes per tourist place ÷ 100 tourists per place per year

10 Spacebus cost per seat to and from orbit

11 Tourists assumed to require half their own weight in supplies

Allowing for other overheads, ground infrastructure costs, partial utilisation, profits and contingencies, the fare for a short visit is likely to be around $20,000.

This is clearly a very preliminary estimate, but it does indicate that the launch cost of the space hotel itself is relatively unimportant and that the cost of a visit is dominated by the cost of the spaceplanes used to transport the tourists and supplies. This is because spaceplanes have to fly far more frequently. Each space station has to be launched once only and will then have a life of more than 20 years. If there are two supply flights per week, for example, spaceplanes will have to make approximately two thousand flights to each space station in that 20-year period.

An analogy can be made with a highly utilised cabin for skiers, placed high on a mountain by a large helicopter crane, with small helicopters used for daily flights carrying skiers, cabin staff, and supplies. Over a period of years, the total cost of the small helicopter flights will far exceed that of the cabin and the large helicopter flight.

The cost per flight of heavy lift vehicles is therefore less important than the cost per flight of spaceplanes. Existing large launchers, such as the Space Shuttle or Ariane 5, could if necessary be used to transport hotel modules, although their cost could be greatly reduced by applying technology from the mature spaceplane.

The enabling development for large new space businesses is therefore the mature spaceplane. Low-cost space stations will follow naturally, and existing heavy lift vehicles could be used if necessary. As soon as mature spaceplanes are more widely recognised as a realistic prospect, we can expect aircraft manufacturers to compete to develop ever more mature space stations and space hotels. We can also expect progress to be rapid, as the engineering fundamentals were sorted out several decades ago.

CHAPTER 7 - THE SPACEPLANE SPACE AGE

We have become so used to the idea that space travel is expensive, risky, suitable only for the super fit - and of necessity a government monopoly - that it is difficult to imagine 'everyday' space activity. But that is precisely what will happen as the result of spaceplane development. In this chapter we will consider what space activities will be like a few years after spaceplanes have reached maturity.

Space science at present is severely limited by cost. Science satellites cost typically hundreds of millions of U.S. dollars, and many experiments that scientists would love to perform cannot be afforded. Selecting instruments for new spacecraft is highly competitive, and many world-class proposals are rejected. Simple 'what if?' ideas rarely get through.

Spaceplanes will reduce the cost of science in space to a level comparable with that in Antarctica, where the cost of access hardly limits the experiments that are carried out. Imagine what science in Antarctica would be like if the only access were by converted ballistic missile. Our robot explorers might now be on the point of discovering penguins!

Observation from space has played a vital part in environmental science, which is the key to understanding human impact on our home planet. It offers the advantage of being able to observe large portions of the atmosphere, oceans, or solid surface at one time. Large environmental research satellites cost up to $2 billion, and very few can be afforded. Low-cost access to orbit will dramatically increase the quantity and quality of the observations that can be afforded. A realistic aim is to have several large observatories in different orbits, some manned, others unmanned.

Astronomy is another discipline that has been transformed by observations from space but, as with environmental science, the experiments have been severely restricted by the cost of access to orbit. With this restriction all but removed, it will be possible to send sophisticated probes to every planet and every large planetary moon in the solar system (more than 50 bodies in all) and return samples.

Much larger, and hence more sensitive, instruments will be feasible. Gamma Ray, X-Ray, ultra-violet, optical, millimetric, infrared, and radio telescopes can be far larger, free from the distortion of gravity, and with assembly in space made practical by low transportation cost. The cost of smaller instruments will be reduced to close to present earthbound ones, which are between 10 and 100 times less expensive than those in satellites today. The lead time will be reduced from typically ten years to two or three. Astronomers will be able to visit their instruments and modify them as appropriate. It will be possible to service unmanned instruments in orbit. There will be astronomical observatories in orbit, specially equipped for visiting amateur astronomers.

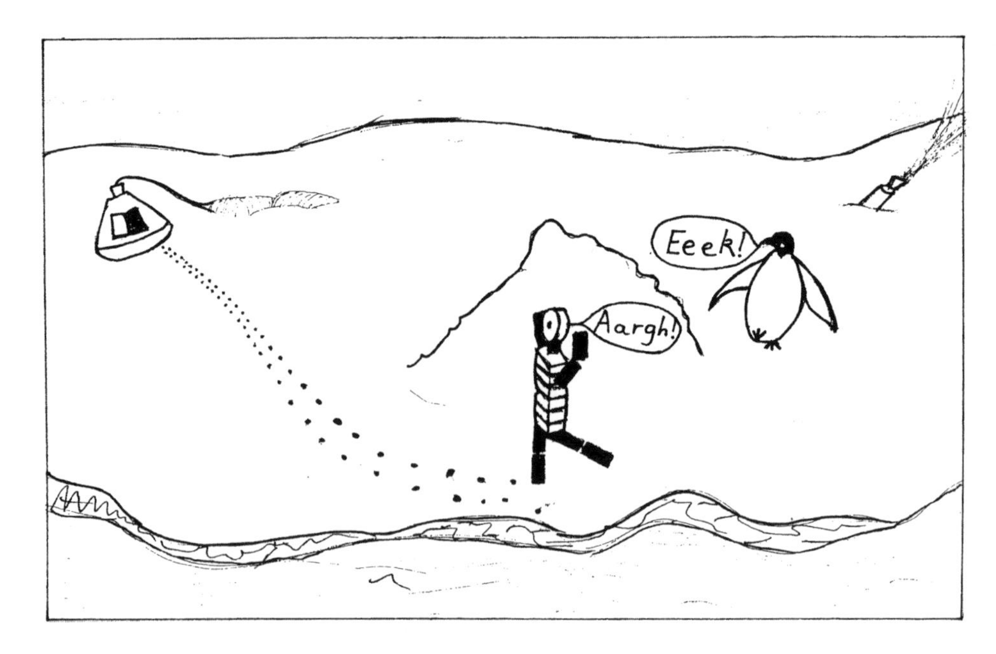

... Imagine what science in Antarctica would be like if the only access were by converted ballistic missile. Our robot explorers might now be on the point of discovering penguins! ...

Factories in orbit will provide near vacuum, near zero gravity, and low-cost power. There have been numerous proposals for manufacturing products under these conditions. The cost of transportation has so far precluded all but very small pilot schemes, although significant research has been carried out. Spaceplanes will remove a large part of the cost barrier. The big unknowns are precisely what products will be made and the size of the market for such products. So far, there have been few if any convincing business plans for manufacture in orbit, even at greatly reduced cost of transport.

Perhaps costume jewellery will be the first commercially significant product to be made in space. Some orbital blacksmith or alchemist is sure to come up with a metallic mixture or crystal of striking appearance that cannot be made easily on Earth, where the convection induced by gravity distorts the solidification and crystallisation processes. This may seem trivial to 'serious' users of space, but one objective of building up demand is to lower the cost for everyone involved, much as computer games helped to spread the use of personal computers and lower their cost.

Large satellites for collecting solar power and transmitting it to Earth have been the subject of much study. Their potential is vast indeed. The energy from the Sun reaching the Earth in just one day equals that in all the known fossil fuel reserves. A satellite of just 250 km diameter could supply all present energy needs, assuming 10% overall efficiency.

Many engineering problems remain unsolved, especially those concerning the transmission of the power to Earth. The high cost of transport to space has so far prevented even small pilot schemes. Not one light bulb on Earth has yet been lit using power from space. Spaceplanes will enable the construction of significant research satellites for solar power collection, although we do not yet know the extent to which our energy needs will be met in this way. However, it is clearly prudent to do research on such future energy options for the human race as soon as the cost permits.

More than 400 people have now been to space since Yuri Gagarin became the first in 1961. Most say that it was a fascinating experience: many go as far as to say transforming. Astronauts rarely tire of looking down at the ever-changing views of Earth, of playing around in zero-g, or looking out at space.

Many ordinary people would love to take a visit to space as soon as it becomes safe and affordable. Dennis Tito became the first space tourist in 2001, when he paid $20 million for a visit to the International Space Station. The great public interest in his visit indicates the latent demand for space tourism. Mature spaceplanes offer the potential of meeting this demand.

If a spaceplane accelerates to just short of orbital velocity, it can then glide half way around the Earth, offering the potential for very fast air travel. A flight from Sydney to London would take about 75 minutes flying time [10]. However, this would probably require a more economical vehicle than would space tourism, which is likely to be commercially viable at a higher fare than would be acceptable for a fast long-distance flight. A visit to space would be marketed initially as a once-in-a-lifetime experience, for which people would pay more than for a fast flight half way round the world.

The long-range sub-orbital airliner would also have to meet more demanding noise and environmental requirements because it would have to take off from close to big cities whereas, if necessary, space tourists could fly from a few airports well away from population centres. Moreover, avoiding sonic bangs over land will be easier for space tourism launches and re-entries than for sub-orbital airliner flights, because of the greater freedom for selecting flight paths.

The long-range sub-orbital airliner would therefore have to be more environmentally friendly than a space tourism spaceplane, as well as more economical, and hence of more advanced design.

This overview of the spaceplane age shows that environmental and astronomical research will be transformed and that space tourism is the most promising new commercial use of space. Manufacturing has great potential, but more research is needed before the commercial demand can be assessed. Solar power generation has vast potential but faces stiff engineering and economic challenges. Long-range sub-orbital airliners for high-speed transport will need to be more advanced than orbital ones for space tourism.

Other 'conventional' uses of space will clearly benefit from the reduced cost of access, but the breakthroughs are likely to be in science and space tourism.

Given these benefits of spaceplanes, the main challenge is finding the development funding. Private sector funding will almost certainly be needed, perhaps supplemented by government support. Private investors will want to see a good return on investment. Since space tourism appears to be potentially the largest new commercial use of space, it is worth considering its potential in more detail.

Later chapters will discuss the likely commercial demand for space tourism, and safety aspects. For now, let us assume that you have saved $20,000 for a visit to a space hotel and consider your experience in more detail.

Much of the preparation will be like that for any other holiday, such as finding the time, getting the best deal, finding somewhere to park the car at the airport, securing the house, and arranging for someone to water the plants and look after the dog.

There will, however, be a health check. Astronauts to date have had to be very fit and healthy. This is because only the most basic medical facilities have been available in manned spacecraft, and medical evacuation has been difficult or not possible. Moreover, with taxpayers spending some $140 million per person per Shuttle flight, it would be disastrous for public relations if an experiment were lost because the experimenter became sick.

These factors will apply far less to space tourists. Spaceplanes can be used as affordable space ambulances, and space hotels will be equipped with medical facilities. However, the tour operators, safety authorities, and insurance companies can be expected to impose moderately severe health checks during the pioneering days of passenger spaceflight, possibly comparable to the checks needed for a private pilot licence. These will be relaxed as confidence is gained from the experience of ordinary people coping with spaceflight. Some new problems may emerge while others may turn out to be less severe than expected. After a few years, most people fit for an active holiday on Earth will probably be able to visit a space hotel. People less fit may have to wait a few more years until additional facilities are available for their comfort and safety.

There will also be a training programme, part compulsory and part optional. The compulsory training will include the use of emergency equipment. This will include a pressure suit that will probably be needed for the flight to the space hotel, and the emergency survival equipment in the hotel itself.

The pressure suit is necessary in case of loss of cabin pressure. The emergency oxygen fitted in airliners would be of little help in the vacuum of space. External pressure is needed to prevent blood from boiling and to enable breathing. The pressure suit will be far simpler than those worn by astronauts or even by high-flying pilots. All it has to do is to keep you alive for a few minutes until the spaceplane can descend back into the atmosphere. There are no requirements for mobility when under pressure, for sophisticated thermal control, or for multi-layer protection, which drive the complexity of existing pressure suits.

The tourist suit will probably be loose fitting and reasonably comfortable to wear, with a visor that closes automatically to seal the suit if cabin pressure is lost. An air supply will then provide the pressure and oxygen required for life support, and the inflated suit will be nearly rigid.

The emergency pressure suit in the hotel will probably be even simpler, consisting of an airtight fabric garment, stored in emergency lockers, that you zip yourself into and await rescue. It will form a sphere of about one metre diameter when under pressure. NASA has already tested such 'rescue balls', designed for possible emergency use on the Space Shuttle.

The optional part of the training will last a few days and will consist of briefings on what to expect and how to make the most of your time in space. Rides will be available to get your body used to the unfamiliar g forces that you will experience. Some of these rides will be like those in theme parks; others will be in aeroplanes flying parabolic trajectories to provide periods of low gravity. The latter are available commercially now in airliners with most of the seats removed to allow space for low-g gymnastics. These provide up to 20 seconds of low-g at a time. Supersonic aeroplanes could give about one minute, and sub-orbital spaceplanes two minutes and more.

One problem not yet completely solved is space sickness. This is related to ordinary travel sickness but is sufficiently different to need special research. It comes on soon after first reaching 'zero-g', and affects a significant proportion of astronauts. Several medications have been developed. One purpose of the theme park and aeroplane rides will be to find the best medication for you, or indeed if you need one at all.

Your seat in the spaceplane will be very like an airliner seat. You will, however, be wearing the emergency pressure suit.

Take-off and initial climb will again be just like in an airliner. You will notice the difference when the Orbiter stage, in which you are seated, separates from the Carrier Aeroplane. You will see the latter disappearing beneath you. When the rocket engines are then set to full throttle, you will hear a louder noise and feel a strong push on your back as the acceleration increases.

The higher the allowable peak acceleration during the ascent to orbit, the less is the propellant needed. It is not a very steep gain, so the acceleration will probably be limited to what most people are comfortable with, possibly around 2g.

As you leave the effective atmosphere you will see the sky turning dark with bright stars even in daytime. At this point you will be able to see the ground for a distance of several hundred kilometres, and the curvature of the Earth.

After about six minutes of acceleration in the Orbiter, the rocket engines will stop, and the noise and vibration will all but vanish. You will feel weightless for the first time. Soon afterwards, the hotel will come into view and, if you are sitting at an appropriate window, you will be able to watch the docking manoeuvre. This will be not unlike a ship berthing at a pier except that it will be in three dimensions rather than two, and that deck-hands attaching lines will be replaced by robot arms.

A few minutes later you will float weightless down the aisle, using 'banisters' to guide you, through the cabin door and the air-lock passage and into the hotel.

There will be two sections to the hotel, one gently rotating and the other still. The rotating section will provide a low effective g level near its rim, probably about one sixth g, so that eating, sleeping, keeping clean, going to the toilet, and other everyday activities are comfortable and do not require special equipment or training. This artificial gravity should also help to quell space sickness.

The non-rotating section will house the viewing ports and the zero-g gymnasium, sports hall, and swimming pool. The view of the Earth is continuously changing as the hotel orbits every 90 minutes. There is a sunset and sunrise every orbit. Features that cannot be picked up even from high flying aircraft become apparent, such as giant eddy currents in the oceans hundreds of miles across, complete cyclones, geological fault lines, and meteor craters. You will be able to identify your own house through a telescope, provided the sky is not cloudy at the time and the orbit passes within viewing distance.

On the dark side of the Earth, you will see a dozen or so lightning flashes per second. The aurorae near the magnetic poles can be seen from above as rippling cones of lights.

Telescopes will enable you to see the Sun, Moon, planets, stars and galaxies far more clearly than from down here, where the atmosphere distorts the image.

In the zero-g gym you will be able to fit wings to your arms and tails to your feet and literally fly like a bird, looking like a medieval or renaissance birdman, Figure 7.1. Your muscles will be strong enough to propel you through the air in the absence of apparent gravity. Your motion will, however, be different from that of a bird down here, because of this lack of apparent gravity. If you stop flapping or twisting your wings or tail, you will carry on in a straight line (relative to the gymnasium), rather than falling towards the ground.

An interesting experiment will be to see how real birds cope with this situation - especially penguins! After some practice, you may end up being able to fly better than eagles. Human flying races will be one of the many new sports made possible by weightlessness.

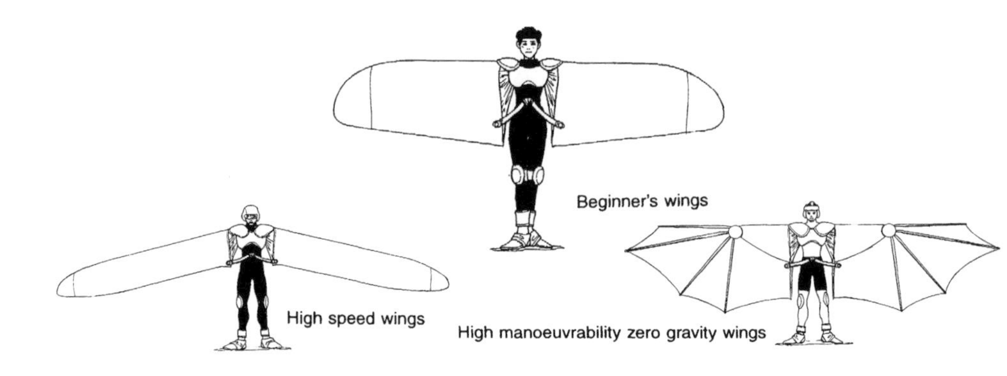

Figure 7.1 Wings for Flight in Low g [From Reference 21]

Low-g swimming will be another new experience, see Figure 7.2. The pool will consist of a large cylindrical drum rotating slowly about its main axis, like a huge washing machine drum. It will contain enough water to provide a metre or so depth around the rim, which is where the water will go due to the rotation. Perhaps surprisingly, you will sink to the same depth as in a pool on Earth, with just your head showing. This is because your apparent weight and the hydrostatic pressure that provides the buoyancy force both decrease by the same fraction as the apparent g is reduced.

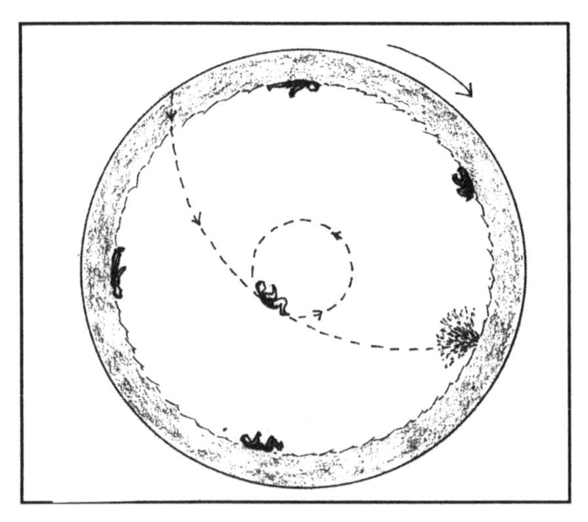

Figure 7.2 Low-g Swimming Pool

The absolute force pulling you down will be far less (assuming that the drum is rotating slowly) and you will be able to kick off from the 'bottom' and land on the water at the 'ceiling'. Your trajectory will be straight relative to non-rotating hotel axes, but curved relative to the drum, because of the rotation. With small wings attached to your arms, you will be able to take off and fly like a flying fish.

There will be optional lectures on geography, Earth science, and astronomy, supported by superb views of the objects in question.

You will return to Earth in the type of vehicle that you flew up in. The return flight will be not unlike the ascent, except that there will be no Carrier Aeroplane, and that the acceleration will be replaced by deceleration. During the peak of re-entry heating, all you will see out of the window will be a hot glow.

Your visit to a space hotel will be even more fascinating than astronaut flights to date, because of the features provided for your entertainment and education.

The remaining chapters explain how such visits to space hotels should be affordable by middle-income people within about fifteen years.

PART 3

TIMESCALE

CHAPTER 8 - TECHNICAL FEASIBILITY

The fundamental engineering problem in designing an aeroplane that can fly to orbit is the high propellant weight that is required. Using late 1950s rocket motors in a single vehicle, about 96% of the launch weight would have to be propellant. This would leave an empty weight of only 4%, which is not feasible, even with materials being researched at present. Even a simple beer can has an empty weight of around 5% of its full weight. This chapter explains how this problem was solved and how by the 1960s a two-stage spaceplane would have been feasible.

A good place to start is with the first vehicle capable of reaching space (but not orbital velocity): the V-2. There were three main sections to the V-2 (Figure 8.1).

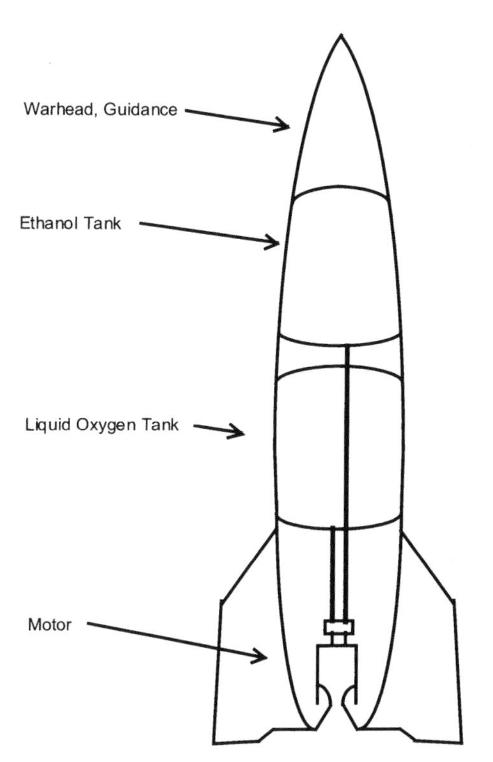

Figure 8.1 Layout of V-2 (Not to Scale)

The head contained the warhead, guidance, and other equipment. The middle section contained two propellant tanks, one for the fuel (ethanol) and one for the oxidiser (liquid oxygen). The rear section contained the rocket engine and the fins. The ethanol was diluted with water to reduce the combustion temperature in the rocket engine to a level acceptable for the materials then available. The propellant mass was 69% of the take-off mass, far less than that required to reach orbit.

A typical flight is shown in Figure 8.2. It started with a period under rocket power of 65 seconds. At the time that the engines shut down, the V-2 was at a speed of 1.6 km per second at a height of 22 km, which is about 5.4 times the local speed of sound. It then coasted unpowered to a maximum height of 85 km, eventually hitting the ground up to 300 km from the launch point.

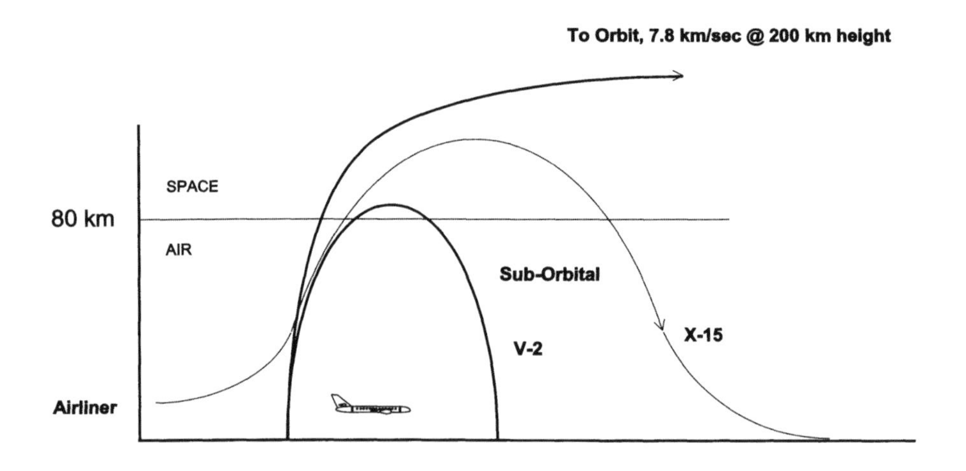

Figure 8.2 Launch Trajectories
The V-2 and X-15 could reach space height,
but their maximum speed was about one fifth of that needed to stay in orbit.

The V-2 pioneered the essential technology used by all subsequent ballistic missiles and launch vehicles: large rocket motors, supersonic aerodynamics, re-entry, and autopilots for guidance and control.

The V-2 followed what is defined as a sub-orbital trajectory, i.e., one that reaches space height but not sufficient velocity to stay up as a satellite. Also shown in Figure 8.2 is a satellite launch trajectory that starts like that of the V-2 but carries on to a 200 km high orbit, requiring a speed some five times faster than the V-2.

The V-2 was, of course, inherently expendable. The first reusable vehicle capable of reaching space height was the North American X-15 experimental spaceplane, Figure 8.3. It was roughly the same size as the V-2 and had broadly comparable maximum speed and height. It was, however, piloted, and was launched from a modified B-52 bomber. It then climbed under rocket power less steeply than the V-2 and landed back on one of the dry lakes at Edwards Air Force Base, Figure 8.2. It made 199 flights between 1959 and 1968, several of which were to space height. During the early 1960s it demonstrated the technology for a reusable launch vehicle.

Figure 8.3 The X-15 Rocket-Powered Experimental Aeroplane, Carried by a B52 [NASA]

The V-2 and the X-15 could achieve roughly one-fifth of the velocity needed to stay up like a satellite. Because of residual atmospheric drag, even in so-called space, an orbital height of 200 km or more is needed for a long-life orbit. Even at this height, occasional 'nudging' by another spacecraft or by small on-board rocket engines is required. Orbits up to a few hundred km above the Earth are often called Low Earth Orbit, or LEO. To reach orbit at a height of 200 km, a velocity of 7.8 km/sec is required, as derived in Appendix 2, resulting in a time to circle the Earth of just under 90 minutes.

As the height is increased, the orbital velocity decreases and the orbital period becomes longer. At a height of around 36,000 km, the period is 24 hours. Thus a satellite in an easterly orbit at that height will remain above the same point on the equator. This so-called geostationary orbit is where most communication satellites are located. This, of course, is why domestic satellite dishes point towards 36,000 km above the equator, usually close to a southerly direction (in the northern hemisphere) but sometimes up to 30 degrees east or west, depending on the location of the relevant satellite.

In order to launch a satellite, the launch vehicle must first carry it clear of the effective atmosphere, and then accelerate it to orbital velocity, Figure 8.2. As well as imparting a velocity of 7.8 km/sec (for a 200-km orbit) the launch vehicle engines have to do work against air resistance and gravity during the ascent. These losses usually add up to an equivalent of about 1.5 km/sec in lost velocity. Thus, the total 'ideal velocity' of the launcher must be about 9.3 km/sec. The equivalent value for the V-2 was 2.4 km/sec. Since the energy required is roughly proportional to the square of the velocity, short range sub-orbital flight is clearly less demanding technically than orbital flight. (Long range sub-orbital flight such as from England to Australia involves acceleration to close to satellite speed, and is very nearly as demanding as orbital flight.)

This is a simplification of the great complexity of calculating ascent trajectories, but is nonetheless adequate for considering the basics of launching satellites.

The high propellant consumption of rocket engines means that, in order to achieve orbital velocity, a high fraction of the launch mass has to be propellant. Figure 8.4 shows that using kerosene fuel (and liquid oxygen) as propellants in a single stage to orbit launcher using modern motors, 94% of the launch mass would have to be propellant.

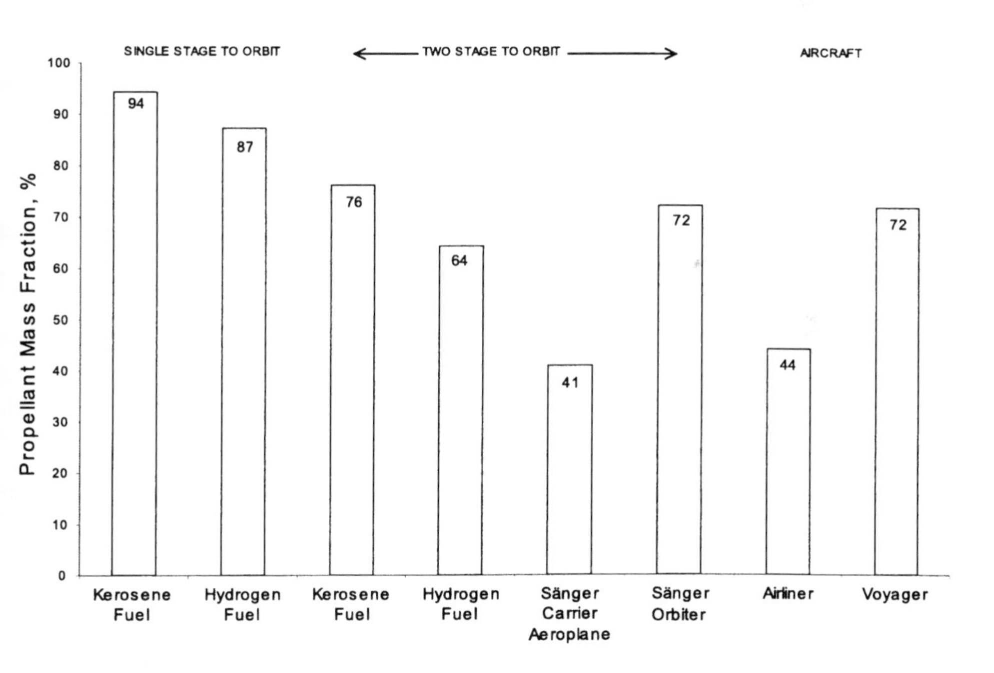

Figure 8.4 Stage Propellant Mass Fractions
Two-stage spaceplanes using hydrogen fuel require a propellant mass fraction no higher than that achieved in long-range aeroplanes.

This is an improvement on the 96% that would be needed with 1950s rocket engines, but still leaves only 6% for everything else, i.e., inert mass (mainly tanks, structure, engines, and equipment) and payload. This is not feasible for a reusable launch vehicle. Some of the most structurally efficient expendable launchers, such as Atlas, Blue Streak and Black Arrow, have achieved inert mass fractions of around 7%. Even this remarkably low figure is too high for a useful load to be carried to orbit, and leaves nothing in hand for the recovery equipment. The equations used to derive these propellant fractions are derived in Appendix 2.

Practical rocket-powered launch vehicles are therefore capable of accelerating to less than orbital velocity: as mentioned earlier, the V-2 and X-15 could reach about one fifth of that velocity. The maximum practical velocity increment for a vehicle is determined by the state of technology at the time, such as specific impulse and structure weight fraction: it does not depend much on the size of the vehicle. The solution has therefore been to use staging. A large vehicle (usually called the lower, or booster, or carrier aeroplane, stage) carries a smaller one, which is released at a significant fraction of orbital velocity when the rocket propellant in the lower stage has been consumed. The small vehicle may in turn carry a smaller one and so on until orbital velocity is achieved. The stage that reaches orbit is sometimes called the orbiter.

Some early launchers had up to five stages, but present-day launchers usually have two stages to reach low Earth orbit, often assisted by strap-on boosters, and a third stage for geostationary orbit.

Using the same kerosene and liquid oxygen propellant combination, and assuming that each stage contributes half the ideal velocity, Figure 8.4 shows that each stage of a two-stage launcher needs a propellant fraction of 76%, leaving 24% for inert mass and payload. If each stage has an inert mass fraction of 10% then each stage has a payload fraction of 14%. The upper stage counts as the payload of the lower stage and therefore has a launch mass of 14% of the launch mass. The real payload (i.e., the satellite) in turn has a mass 14% of the upper stage launch mass. Thus the payload mass is 14% of 14%, which is only 2%, of the overall launch mass. Figure 8.5 shows the top-level mass breakdown of a two-stage vehicle with these mass fractions.

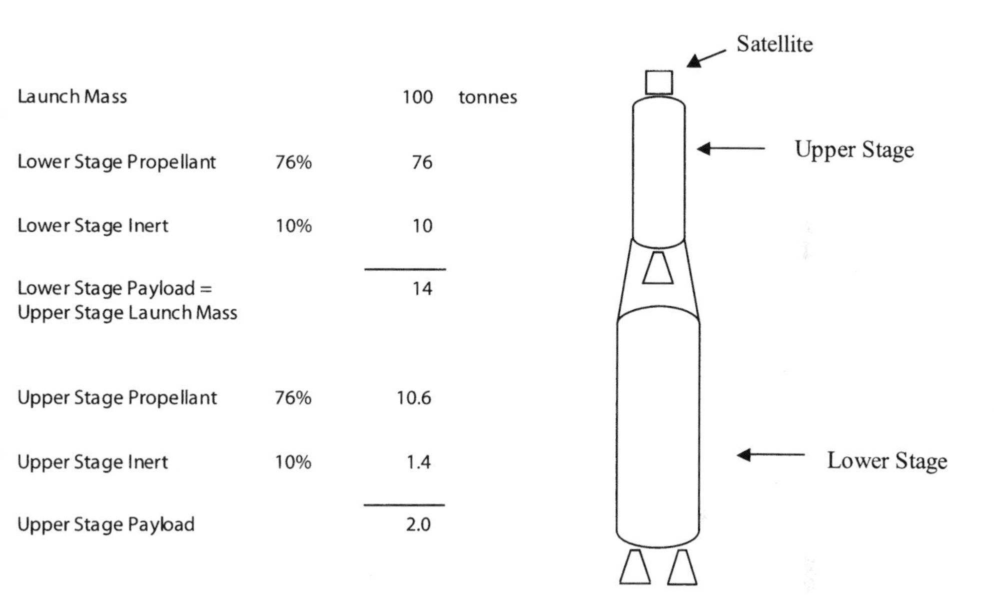

Figure 8.5 The Principles of Staging
With existing technology, launchers with more than one stage are needed to launch satellites.

These numbers are illustrative, but serve to show why launchers use more than one stage and why they have a very low payload as a fraction of launch mass. The numbers also show why early launchers were expendable. There was insufficient weight in hand for recovery equipment, such as wings, tail, flying controls, landing gear, thermal protection, and (optionally) cockpit and crew.

An earthbound analogy with staging is a base camp used in mountaineering. Suppose for the sake of argument that you want to climb a mountain 8,000 metres high, that you need 20 kg of supplies to climb each 4,000 metres, and that you can carry a maximum load of 25 kg. The weight of the supplies you need to climb the mountain is then 40 kg, which is greater than the load you can carry. You can, however, carry a 5 kg surplus to a point half way up. The solution is to persuade three of your friends each to carry 5 kg of supplies half way. You accompany them and collect the 15 kg of supplies that they carried for you. Together with the 5 kg you carried for yourself, you have enough to get to the top.

This is of course a gross oversimplification of the logistics of mountaineering, but it shows how a staged effort can achieve what a solo one cannot, at the expense of size and complexity. In the mountaineering example, the number of people had to be increased four times to go twice as high. It would have to be increased sixteen times to go three times as high, and so on.

Another solution is to use propellants with higher energy. The practicable combination with the highest energy is liquid hydrogen and liquid oxygen. There are propellant combinations with somewhat higher energy, but these have major problems such as high cost and/or toxicity. Figure 8.4 shows that, with hydrogen fuel (and liquid oxygen), the propellant fraction on a single-stage vehicle is about 87%, leaving 13% for inert mass and payload. This is high enough to have tempted several companies to propose single-stage projects using hydrogen fuel, the latest of which was the Lockheed-Martin Venture Star, mentioned earlier.

With two stages and hydrogen fuel, the propellant fraction on each stage is down to 64%, allowing 36% on each stage for inert mass and payload. This is well into the practicable region. Hydrogen fuel was first flown successfully on the Centaur upper stage in 1963, and many of the 1960s spaceplane designs had two stages and hydrogen fuel. The designers at the time considered their projects to be feasible and, with hindsight, many of them were right.

Another way of reducing the propellant fraction is to use air-breathing engines. Two broad categories of propulsion system can be used on a launch vehicle: air-breathing and rocket, as shown in Figure 8.6. As the name implies, air-breathing engines take in air from the atmosphere. This eliminates the need to carry the oxidiser on board, thereby greatly reducing propellant consumption. One type of air-breathing engine is the familiar jet. Air enters the engine and is compressed by two sets of blades. One set is mounted on a rotating drum and the other set is mounted on the engine casing. Fuel is injected into a combustion chamber where it burns with the oxygen in the compressed air. The heat so generated causes the air to expand and to be forced out through a nozzle as a jet of high-speed exhaust gas. On the way, it passes through a turbine that drives the compressor. The thrust is the reaction from speeding up the air flow.

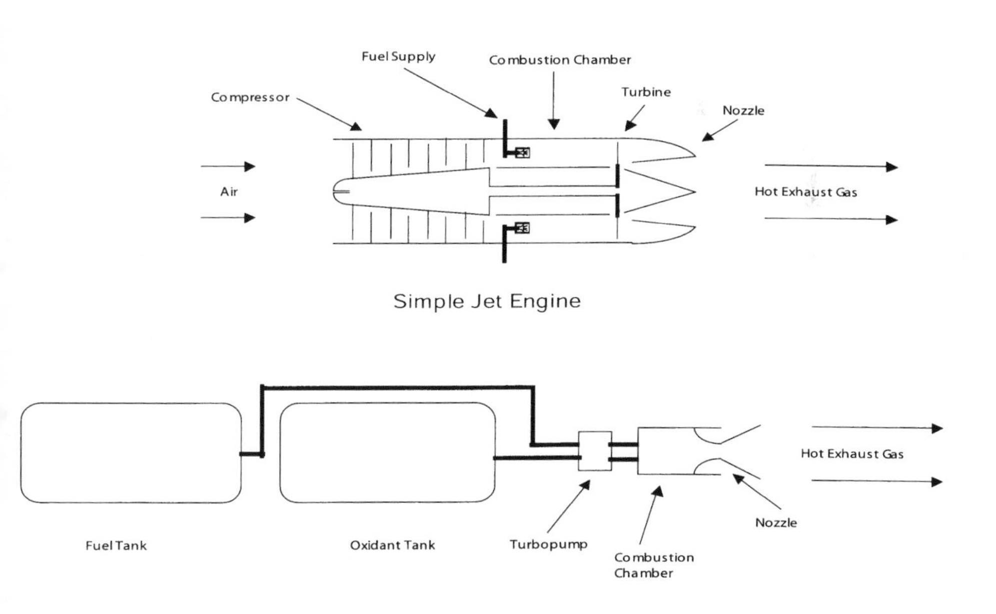

Figure 8.6 Jet and Rocket Engines

A liquid-fuelled rocket engine has much in common with a jet engine. The main differences are that the vehicle carries its own oxidiser, often liquid oxygen, and that this oxidiser is pumped into the combustion chamber as a liquid rather than as a gas. The turbine driving the pump is a separate unit, rather than being driven by the main exhaust stream. The temperature in a jet engine combustion chamber is usually lower because the nitrogen in the air absorbs much of the heat.

Because they have to carry their own oxidiser, rocket engines have a far higher rate of propellant consumption than jet engines. A modern rocket engine using liquid oxygen and kerosene propellants will produce about 330 kg of thrust for each kg of propellant consumed per second. The so-called specific impulse is then 330 sec. An airliner jet engine at subsonic speed uses about 0.6 kg of fuel per hour for each kg of thrust. Converting the units, this is equivalent to 6000 kg of thrust for each kg of fuel consumed per second, which is 18 times more efficient than the rocket engine.

Figure 8.7 Lockheed X-7 Test Missile, 1950s [Lockheed Martin]

The X-7 was used in the 1950s to flight-test high-speed ramjet engines, and achieved a maximum speed of Mach 4.3, the highest yet by an air-breathing engine. Even this is only about one sixth of orbital velocity.

On the other hand, conventional jet engines weigh ten to twenty times more than rocket engines of comparable thrust. Moreover, as already mentioned, the fastest speed yet achieved by an air-breathing engine is the Mach 4.3 of the Lockheed X-7 test missile of the early 1950s, Figure 8.7, which is only one sixth of orbital velocity.

Existing air-breathing engines are too slow and heavy for a single-stage launcher, although they could be used on the carrier aeroplane of a two-stage vehicle. Even the advanced air-breathing engines proposed for Sänger, capable of Mach 7, would be too slow and heavy for a single-stage vehicle, which is why Sänger has two stages. As shown in Figure 8.4, the Sänger Carrier Aeroplane has a propellant fraction of 41% and the Orbiter 72%.

More advanced air-breathing engines are needed for a practicable single-stage-to-orbit vehicle. One proposed solution uses a combined jet and rocket engine. Air entering the intake is chilled in a heat exchanger, using the hydrogen fuel as coolant. This reduces the size and weight of the compressor needed to force the air into the combustion chamber. At a speed of about Mach 5, the drag caused by slowing the air down in the intake becomes so high that a pure rocket engine becomes more efficient. At this speed, the air flow is cut off and is replaced by liquid oxygen carried on board the spaceplane. The same combustion chamber and nozzle are used for both the air breathing and rocket portions of the flight.

This is the basis of the Reaction Engines Limited SABRE engine proposed for their SKYLON single-stage spaceplane, shown in Figure 8.8, which is derived from the British Aerospace HOTOL project of the 1980s. The maximum air-breathing speed of Mach 5 leaves the rocket engine with a high speed range over which to accelerate the vehicle, requiring an overall propellant mass fraction of 79%, which is high by present robust aeroplane standards but by no means unfeasible.

Figure 8.8 The Proposed SKYLON Single Stage to Orbit Spaceplane [Reaction Engines]
Skylon uses a combined air-breathing and rocket engine to reduce the propellant mass fraction needed to reach orbit with a single stage spaceplane.

To reduce the propellant fraction even further, air-breathing engines faster than Mach 5 are required. To overcome the problem of high drag in the intake, the air must be slowed down less, which means combustion at supersonic speed. The resulting engine is called a supersonic combustion ramjet, or scramjet. The X-43 scramjet unpiloted test vehicle, Figure 8.9, is designed for Mach 7, with the potential of Mach 10 later. Its first launch, in 2001, was not successful because of a booster failure.

Figure 8.9 The X-43 Scramjet Research Vehicle [NASA]
The 3.7 metres long X-43 is designed to test supersonic combustion ramjets in flight. These offer the long-term prospect of an aeroplane-like single stage to orbit vehicle.

In principle, such an engine can be very simple, with no moving parts. The air is compressed simply by being slowed down, as much of its kinetic energy is converted into pressure energy. A compressor and the turbine to drive it are not needed. Unfortunately, such engines do not work at all at low speeds, and additional propulsion is needed for take-off and early acceleration. Moreover, a variable intake and nozzle are needed for such an engine to be efficient over a wide speed range.

There are several technical challenges to overcome before a practical scramjet can be developed. The air becomes very hot as it is slowed down, as some of its kinetic energy is converted into heat. The engine materials must therefore either be cooled or capable of withstanding very high temperatures. Another challenge is that the energy in the high-speed air stream becomes comparable with the energy added by burning the fuel. Losses in the intake and nozzle can then easily become greater than the fuel energy, leading to a net drag rather than the required thrust. It is therefore necessary to achieve high efficiencies in the intake and nozzle.

There have been numerous proposals for other types of high-speed air-breathing and combined air-breathing/rocket engines. However, as will be discussed later, the first successful mature spaceplane will probably use developments of existing engines and have two stages, leaving advanced air-breathing engines for second-generation developments.

For a sub-orbital spaceplane, the propellant fraction required is well within present feasibility. The sub-orbital V-2 had a propellant fraction of 69%, and the X-15 around 58%. The X-15 used more advanced engines than the V-2, and was air-launched. With modern rocket engines, the same performance could be obtained with a propellant fraction of around 50%.

Useful insights into the challenge of spaceplane design can be gained by comparison with long-range aeroplanes, because both share the requirement for a high propellant mass fraction. (In aeroplanes, this is more usually called the fuel weight fraction.) Figure 8.4 shows that the fuel weight of an airliner on a long-range flight is typically 44% of the take-off weight. Tanker aeroplanes used for in-flight refuelling achieve fuel weight fractions up to about 56%. The longest-range aeroplane to date, and the one with the highest fuel weight fraction, is the Rutan Voyager, Figure 8.10, which was designed specifically to fly round the world non-stop without in-flight refuelling. This it did with great success in 1986. It was a one-off design and not practical for everyday use. Given a commercial incentive, however, a vehicle with such a high fuel fraction could undoubtedly be developed to a mature stage.

Figure 8.10 Rutan Voyager [Scaled Composites]
The Rutan Voyager could take off with 72% of its weight as fuel.
This is a record for an aeroplane, and is more than adequate for a two-stage spaceplane.

The Voyager had an advanced structure that weighed less than 10% of the take-off weight. As a result, it could carry 72% of its take-off weight as fuel. By coincidence, this figure is similar to that of the Sänger orbiter, which is good evidence for the feasibility of Sänger. For the actual record, the fuel consumed was 70% of take-off weight.

The structural efficiency of a spaceplane with a 70% fuel fraction therefore has to be very roughly comparable to that of an aeroplane that can fly around the world once. We can now find the 'equivalent aeroplane range' of the single-stage and two-stage launchers shown in Figure 8.4 (i.e., the distance that an efficient aeroplane would fly with the same propellant mass fraction), as in the following table:

Table 8.1

AEROPLANE CONFIGURATION	FUEL FRACTION %	RANGE/EARTH CIRCUMFERENCE
Voyager	70.1	1.00
Single-Stage, Kerosene Fuel	94.3	2.38
Single-Stage, Hydrogen Fuel	87.3	1.71
Two-Stage, Kerosene Fuel	76.2	1.19
Two-Stage, Hydrogen Fuel	64.3	0.85

The propellant fractions are taken from Figure 8.4. The range, measured in Earth circumferences, is scaled from Voyager by a method explained in Appendix 2. Thus, very roughly speaking, the difficulty of building a single-stage launcher with kerosene fuel is comparable to that of an aeroplane capable of flying 2.4 times round the Earth, which is well beyond present aeronautical capability. Even with hydrogen fuel, a range of 1.7 times round the Earth is required, which again is well beyond present capability.

For the two-stage vehicles, the required equivalent aeroplane range of each stage is half that of the corresponding single-stager, as might be expected. With kerosene fuel, it is somewhat further than Voyager and with hydrogen fuel somewhat less far.

This analogy helps to explain the need for two stages and the advantages of hydrogen fuel. It helps to put spaceplane design in an aeronautical perspective. Any respectable aeroplane designer would soon come to appreciate that an aeroplane that could fly twice round the Earth was not a practical proposition, except perhaps with non-existent nuclear engines or very slowly using solar power. Designers of single-stage-to-orbit launch vehicles have to solve a comparable problem, unless they use advanced air-breathing engines to reduce the propellant fraction.

Some of the early attempts at trans-Atlantic commercial air travel used a two-stage approach. Two examples are shown in Figure 8.11. Both used a Short Empire flying boat. The top illustration shows flight refuelling over Southampton Water in England. The flying boat then went on to cross the Atlantic non-stop in 1939. Flight refuelling is not strictly a two-stage operation in the sense in which we have been using it, but it fulfils a similar purpose. It has never been used for commercial flying, but is widely used to extend the range of military aircraft.

The lower illustration shows a small mailplane, the Short Mercury, being carried by a modified Empire flying boat. Such assisted take-off enabled the Mercury to carry a heavier fuel load and thereby fly further. The Mercury also flew across the Atlantic non-stop, in 1938.

World War II cut short these experiments. Otherwise, flight refuelling or two-stage aeroplanes might have been commercially successful for carrying trans-Atlantic mail. If flying across the Atlantic required the same fuel fraction as flying to orbit, it is probably that two-stage airliners would have been used for early passenger flights, until the technology for a single-stage vehicle could have been developed.

Figure 8.11 Early Experiments with Flight Refuelling and Two-Stage Aeroplanes
[Flight Refuelling Ltd and Short Brothers plc]
Above: Handley Page Harrow Refuelling Short Empire Flying Boat, 1939.
Below: Short-Mayo Composite, 1938.
The Short-Mayo composite demonstrated the prospect of safe and reliable two-stage operation.

Summarising this consideration of the propellant mass fraction required to reach orbit: two-stage spaceplanes using hydrogen fuel have been feasible since the late 1960s, and a practical single-stage launcher requires advanced air-breathing engines.

Other technologies not found on conventional airliners are required for a spaceplane. The more important are: large rocket engines; thermal protection; hot structures; reaction controls; systems for guidance, navigation and communication in space; lubricants for use in space; the aerodynamics of large delta-winged aeroplanes; lightweight propellant tanks; and protection against acoustic fatigue. These technologies have been developed for Concorde, Space Shuttle, X-15, Buran (the Russian equivalent to the Space Shuttle that flew only once), and various large satellites; and are adequate for a prototype two-stage spaceplane but not yet for a mature one.

CHAPTER 9 - SAFETY

Safety will be a key issue with passenger-carrying spaceplanes. To put this in perspective, Figure 9.1 compares the fatal accident rate of various types of flying, rounded off to the nearest order of magnitude for the sake of broad-brush comparison.

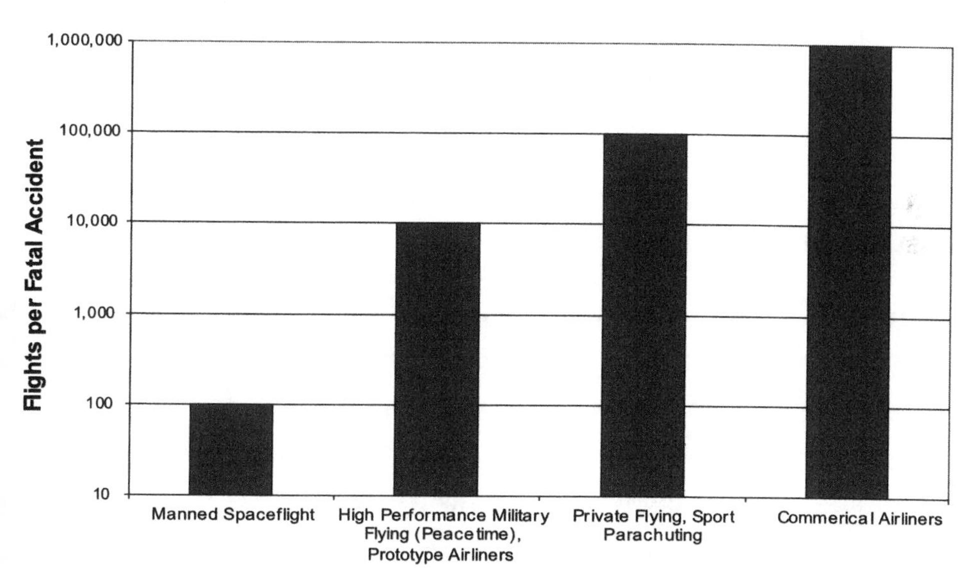

Figure 9.1 Safety Comparisons

Manned spaceflight is at present approximately 10,000 times more risky than airline flying.

By the end of the year 2001, the United States had made 137 manned space flights with one fatal accident - the Challenger Space Shuttle crash in 1986. The Soviet Union and Russia had made 94 manned space flights with two fatal accidents. The average number of flights per fatal accident is therefore around 100.

Peacetime training in high performance military aircraft is about 100 times less hazardous, with a rate of 10,000 flights per fatal accident. To be effective, combat pilots have to approach their own limits and those of their aircraft. By contrast, in commercial flying pilots are encouraged to keep as far away as possible from such limits.

Test flying airliners or business jets has a comparable accident rate. Roughly one in ten new types of airliner or business jet has a fatal crash in its flight test programme. These tend to be 'freak' accidents, caused by a combination of things going wrong. Since each type makes roughly 1000 test flights, the fatal accident rate is about one in 10,000 flights.

Light aeroplane flying has a fatal accident rate of around one in 100,000 flights, and sport parachuting has a comparable accident rate of around one fatality per 100,000 jumps.

Commercial airliners have a fatal accident rate of around one per million flights. Airline pilots are more thoroughly trained and monitored than amateur pilots, and have to go through a selection process and pass a strict medical exam. Given all the things that can go wrong during an airliner flight and the bad weather that they routinely fly through, this low accident rate is an outstanding accomplishment. It is the result of continuous high priority effort since the earliest days of flight.

It is very likely that spaceplanes will have to be as safe, or at least nearly as safe, as airliners before they are allowed to carry passengers, which is some 10,000 times safer than manned spacecraft to date. The Federal Aviation Administration (FAA) in the United States, the forthcoming European Aviation Safety Agency (EASA), and other airworthiness authorities, will almost certainly insist on a rigorous flight-testing programme before issuing a passenger-carrying type certificate. There may be some relaxation during the pioneering stages when the number of passengers is small, but aspiring spaceplane manufacturers should expect to expend a great deal of effort on safety.

In order to assess the feasibility of this safety target, a good place to start is to consider why manned spaceflight has been some 10,000 times more risky than airliner flights.

The overriding cause is the use of throwaway launchers using ballistic missile technology. It is not practicable to aim for high safety with expendable vehicles because their high cost per flight precludes a full flight-testing programme. A new airliner makes typically 1000 test flights before the authorities will allow it to carry fare-paying passengers, and a high-performance research aeroplane typically takes 100 flights to complete its test programme.

An incremental flight-testing programme is followed. Each flight 'pushes the envelope' a little. It is a little faster, or slower, or higher than previous flights, or tests a system nearer to its design limit. If something does not work as planned, there is a proven flight envelope to fall back into.

This can be afforded because the vehicle is fully reusable and the marginal cost items per flight are crew salaries, maintenance, and fuel. By contrast, the marginal cost of an expendable vehicle flight is the cost of a new vehicle, which severely restricts the number of test flights and greatly increases the risk. For example, the first powered flight of the Space Shuttle, following five gliding flights, was to orbit. Each system had to work right first time close to its limits. Such a leap into the unknown would be unthinkable with a new aeroplane. Concorde, for example, made 69 flights before reaching even supersonic speed.

This decision to fly to orbit on the first powered flight was not taken lightly and, due to great skill and dedication, it came off. However, some luck was involved. The first Space Shuttle re-entry showed that the angle of the flying controls needed to trim the vehicle was twice that predicted, and not far short of their maximum angle. Had the error been much greater, the vehicle would have become uncontrollable. This type of problem would normally be observed in an incremental flight-test programme long before it became serious.

Another reason why expendable launchers cannot be made safe is that vehicles off the production line cannot be tested in flight before being used, because they can fly once only. Many complicated systems therefore have to work right first time. By contrast, airliners off the production line usually make two or three verification flights at the manufacturer's airfield, followed by a handover flight to the customer airline, who will generally put the aircraft directly into service.

Yet another reason is that the economics of expendable vehicles leads to a lack of back-up systems and to low design margins. Improving either of these aspects would increase weight and cost, and reduce the size of satellite that could be carried.

Considering these factors, the level of safety on manned space flights that has been achieved with expendable launchers is a remarkable achievement.

Because a spaceplane is in engineering essentials an aeroplane, all of the safety problems due to expendability are avoided. It should therefore be possible with a major effort to achieve safety approaching that of airliners. Incremental flight-testing and pre-delivery acceptance flights can be afforded, and adequate system redundancy and design margins can be built in. There are, however, some new or increased hazards due to high-speed rocket flight, such as the use of high-energy rocket propellants, re-entry heating, and stability and control at high speeds and high angles of incidence. There has been enough experience with rocket aeroplanes over the years to be confident that adequate solutions can be found, although a lengthy programme of analysis and testing will be needed. The hazards of operating in space are considered later.

As a thought experiment to illustrate the potential benefits of spaceplanes in terms of cost and safety, consider a Shuttle Orbiter developed to fly as a fully reusable aeroplane by carrying propellant in its cargo bay and with modifications to allow it to take off down a runway, as shown in Figure 9.2. The required modifications would probably be straightforward. Jet engines would be added to improve the safety of approach and landing, making it possible, for example, to divert to a different runway in case of emergency. The resulting performance would be short range sub-orbital only. The 'Sub-Orbital Shuttle Aeroplane' would take off, climb briefly to space height, and then land back where it started from.

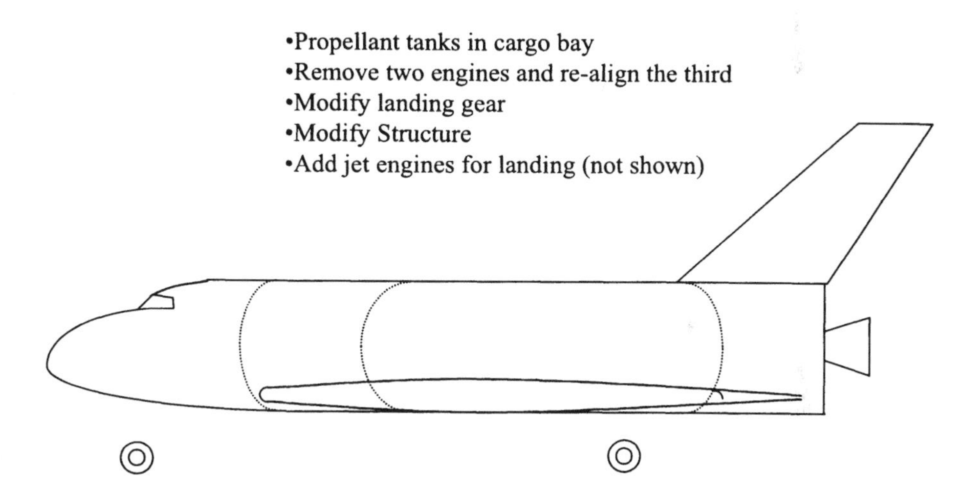

Figure 9.2 Concept Sketch of Shuttle Orbiter Modified for Autonomous Flight
A Space Shuttle Orbiter modified to take off and land unassisted could in principle be made as safe as a conventional aeroplane, because a full flight test programme could be afforded.

This Sub-Orbital Shuttle Aeroplane would be like an enlarged rocket fighter such as the Saunders Roe S.R.53 mentioned earlier.

There is little doubt that, with a thorough flight test programme and detailed design improvements over a period of perhaps ten years, it could approach airliner safety. There is simply no reason why it should not, given the engineering effort to mature the rocket motors and other systems. Such improvement would be expensive, but the low marginal cost per flight would enable such a programme to be afforded if there were a strong commercial incentive.

This thought experiment illustrates the point that the poor safety of the Shuttle compared with airliners is largely because the cost of expendability has precluded an approach to aeroplane-like maturity. Thus, the Shuttle made about 100 flights in its first 20 years, whereas a new airliner makes typically 1000 test flights in its first year of flying. Another contributing factor to the poor safety of the Shuttle is the use of solid rocket boosters. Once lit, these cannot be stopped until all the propellant is burnt. By contrast, liquid-fuelled boosters can be shut down if a problem is detected. Analogously, a domestic coal fire is harder to put out than a liquid-fuelled one.

The idea of such a 'Sub-orbital Shuttle Aeroplane' is somewhat academic because there is little commercial incentive, but it serves to illustrate that spaceplanes are potentially as safe as aeroplanes.

Safety challenges due to operation in space include the potentially disastrous effects of loss of cabin pressure, radiation, and space debris. These hazards affect both spaceplanes and space hotels. The emergency space suits and 'survival balls' mentioned earlier should provide adequate protection in the unlikely event of loss of cabin pressure, either in the spaceplane or in the space hotel.

Radiation from the Sun and more distant bodies is higher in space than on Earth because the atmosphere serves as a shield. The average radiation dose on the surface of the Earth due to natural causes is 2.5 milli Sieverts per year. There is considerable local variation. In the county of Cornwall, England, for example, the average dose is 8 milli Sieverts [22], because of the presence of naturally occurring radon gas. Workers in the nuclear industry are allowed an annual dose approximately ten times the average naturally occurring dose, and the NASA annual limit for astronauts is ten times higher again [23].

Measurements inside orbiting spacecraft indicate that astronauts are exposed to about 0.3 milli Sieverts per day, although the dose varies considerably with time, type of orbit, spacecraft orientation, and with location inside the spacecraft. Roughly speaking, a week in a spacecraft results in a similar dose to a year on Earth.

Far higher doses occur during solar flares, which produce doses from 1000 to 20,000 milli Sieverts. The fatal dose is approximately 3000 milli Sieverts, so solar flares present a major hazard, especially as they are irregular and unpredictable. However, they can be seen erupting from the Sun several minutes before their radiation reaches Earth, which is time enough to move to an on-board radiation shelter, as described later in this chapter. Solar flares with potentially serious radiation are not frequent. No astronaut has yet received a lethal dose in space.

The flight to and from orbit in a spaceplane is of such short duration that little or no protection against radiation should be needed for passengers. Neither of the two present spacefaring people carriers, the Shuttle and Soyuz, has radiation shielding. The highest-flying airliner, Concorde, carries radiation meters so that it can descend to lower altitude in the event of intense radiation from solar flares but, at the time of writing, this has not yet happened.

The spaceplane flight crew will need close monitoring to ensure that they are not exposed to too much radiation. This has recently become an issue with airliner flight crews, who are exposed to about twice the average terrestrial radiation. For many years, no increase in the cancer rate of flight crews was detected. However, recent techniques with higher resolution have provided inconclusive evidence that radiation may be a problem with flight crew. There is at present no obvious solution because fitting cockpits with available shielding materials would be prohibitively heavy.

The situation in space hotels is different from that in spaceplanes in two ways. First, the exposure time is far longer. Crews will almost certainly need protection and passengers probably so. Second, the cost of providing shielding should be manageable because it has to be launched once only and then has an indefinite life in orbit.

While it is too early to describe a detailed solution, the general principles are clear. Space hotels will have layered defences. Their orbits will be beneath the van Allen radiation belt, which provides significant protection, at least until the problem is better understood. Sun-watching instruments will give notice of forthcoming intense radiation, allowing crew and passengers time to retreat to highly shielded 'citadels' where they will be safe until the hazard has subsided or until they can be evacuated back to Earth. Crew living quarters will probably have significant shielding against background radiation and passenger quarters less. Water and fuel may be stored in hollow walls to increase protection.

Space debris is a general term given to the many particles in space that are not recognised astronomical objects. It derives from two main sources: meteorites from the depths of the solar system and the remains of satellites and rockets that have collided or exploded. It is a depressing fact that, in some of the more commonly used orbits, more than half the space debris near Earth is from the latter source. Efforts are now being made to reduce the formation of new debris.

A recent estimate [24] suggests that a permanent space station could expect an impact by a dangerous, one centimetre particle, approximately once every 200 years. This may seem like a low risk but, if there were 100 space hotels in orbit, then on average one would be damaged every two years. Such an impact would puncture a pressure hull and cause a slow leak, or would destroy a piece of equipment.

Some debris is large enough to be detected, and this can in principle be destroyed or deflected by space-based weapons, or avoided by moving the hotel out of the way. Most debris is too small for this defence, and some shielding will probably be needed. Space hotels will be built in modules that in emergency can be self-contained, each with equipment essential for survival. In the event of serious damage to one module, passengers and crew can evacuate to an undamaged one. Radiation shielding should also be effective against debris, so that duplication should not be needed.

Long-duration space flights have shown that months of zero-g can cause bone embrittlement and other medical problems. Tourists on a visit of a few days will not be affected. For longer stays, and for crew, the solution is to provide sufficient artificial gravity in a rotating section of the space hotel to avoid these problems, although the required combination of g and time has yet to be established.

Solutions to the new safety challenges are thus available at concept level. Implementing them will require the same approach as used for managing risk in other transportation systems, i.e., sound engineering practice, minimising exposure to risk, shielding, system redundancy, highly trained staff, adequate regulation, highly professional accident investigation, confidential incident reporting, rapid communication of lessons learned, statistical analysis, and painstaking detailed development; in short, a good safety culture.

There has been enough experience of manned spaceflight to be sure that there are no insurmountable problems. The scale of the engineering effort that will be needed to achieve the required safety is less easy to predict but, if engineering history is a guide, this is more likely to increase fares than to delay the start of space tourism. As soon as the business potential of space tourism becomes widely appreciated, research into space radiation and debris is likely to accelerate.

These safety challenges are all present for a sub-orbital spaceplane but the duration is far less and the problems thereby less severe. A sub-orbital spaceplane is therefore the ideal lead-in to an orbital one.

CHAPTER 10 - MATURITY

Although the prototype of a two-stage spaceplane could be built soon with existing technology, it would fall far short of airliner maturity. It would need extensive maintenance between flights and would have a short life.

The most expensive maintenance item would probably be the rocket motors, which now have lives of between 10 and 100 flights, compared with up to 10,000 for airliner jet engines. A development programme to increase rocket motor life would have much in common with that used to improve jet engines over the years. The problems are similar, i.e., corrosion, creep, erosion, fatigue and wear. The required expertise, facilities, and techniques are also similar. Higher margins, more redundancy, condition warning, and intensive materials research and component development will be required.

The main reason for the short life of rocket motors is that there has never been a strong military or commercial demand for long-life ones. It is therefore relevant to compare the history of jet and rocket engines. Aircraft rocket engines were invented at about the same time as jet engines, as shown in Figure 10.1.

The first turbojet aeroplane was the Heinkel He 178, which flew in August 1939. The second (and arguably the first practical one), the Gloster E.28/39, flew in May 1941. The first jet fighters, the Messerschmitt Me 262 and the Gloster Meteor, entered service in 1944, too late to have much influence on World War II. The Junkers Jumo 004 jet engines in the Me 262 had a life of around ten flights, comparable to the life of present rocket motors.

In 1952, just eight years later and following intensive development of the jet engine, the de Havilland Comet started the first jet airliner service.

The first operational rocket aeroplane was the Messerschmitt Me 163 fighter of World War II, which entered service in 1944. It was by far the fastest combat

Heinkel He 178.
First Jet Aeroplane to fly, 1939

Gloster E.28/39, 1941

Messerschmitt Me-262, operational in 1944

Gloster Meteor, operational in 1944

Messerschmitt Me-163.
The first rocket fighter, operational in 1944

De Havilland Comet.
Started the First jet airliner service in 1952

Figure 10.1 Early Jet and Rocket Aeroplanes

Jet and rocket engines for aeroplanes entered quasi-experimental service in 1944. The first jet airliner service started just eight years later. Rocket engines suitable for passenger spaceplanes could be developed in a comparable timescale, given priority.

aeroplane in that conflict but had such a short flying time that it was ineffective as a weapon system. Its Walter 109-509 rocket engine and associated propellant tanks were notoriously prone to blowing up, and the war ended before the design could reach maturity. This rocket motor achieved a greater number of flights than any subsequent unit designed for aeroplanes. There have been no operational military rocket planes since the Me 163, and the main use of rocket engines since then has been on missiles and expendable launchers, which require but a single firing.

A useful indication of the maturity (and hence the cost per flight) of an aeronautical technology is the cumulative total number of flights that have used that technology. Rocket engines have now flown as many total flights as jet engines had by the mid-1940s.

There have, however been some aeroplane rocket engines since the 1940s. One such was the Thiokol XLR 99, used in the X-15 [25]. The safety requirement for this engine was that "any single malfunction in either engine or propulsion system should not create a condition which could be hazardous to the pilot". Thiokol claimed to have met this requirement and, indeed, no X-15 was seriously put at risk in the air due to a propulsion system failure. Development tests showed that the life averaged around 100 flights. The limiting component was the thrust chamber. Erosion at the throat led eventually to pinhole leaks in the tubes that formed the throat and which contained the cooling propellant. In flight, however, the engine did not last that long. Eight engines were used for 199 flights - an average life of almost 25 flights per engine.

Perhaps the closest yet to a mature operational rocket fighter was the British Saunders Roe SR.53, mentioned earlier. Its de Havilland Spectre rocket motor used technology transferred from Germany when the designer of the Me 163 engine, Helmuth Walter, moved to England after World War II. The Spectre was designed as an operational aircraft powerplant. Had this aeroplane entered service, no doubt its rocket engine would have been developed to have a life and maintenance cost comparable with military jet engines.

The Bristol Siddeley BS 605 rocket motor, developed from the Black Arrow launcher engine, was used operationally to boost the take-off performance of the Blackburn Buccaneer bombers of the South African Air Force. According to a brochure, this engine had a life of 60 firings or 30 minutes total operation between overhauls. The main task at overhaul was changing the catalyst pack. During development, some thrust chambers had a life of more than 100 firings. To increase life and reliability, this engine was run at lower thrust levels than the expendable engine from which it was derived.

The Russian NK-33 engine, designed for the Moon programme, has a life of more than twenty flights, and is planned for the Kistler K-1 reusable launch vehicle.

The Space Shuttle Main Engine was designed to be reusable, but several components have to be removed for maintenance between flights. A progressive series of upgrades continues. The design goal is to operate for 7.4 accumulated hours, equivalent to some 45 flights.

The Rocketdyne RS-83 engine is in the early stages of development for use on reusable launch vehicles. It is designed for a life of 100 flights.

As an indication of the short-life culture prevalent in rocket motor establishments, the durations of firings in a test cell are quoted in seconds. The corresponding times for jet engines are quoted in hours. A new rocket engine makes between 150 and 1500 test firings before the first flight, whereas a new jet engine will undergo some 12,000 endurance cycles before flying [27].

As soon as the potential for large new businesses using spaceplanes becomes more widely appreciated, we can expect the engine companies to compete in developing rocket engines with longer life and lower maintenance cost. As mentioned earlier, it took eight years for jet engines to mature from quasi-experimental military use to the first jet airliner service. A comparable timescale is likely for the development of long-life rocket engines, once there is a strong commercial incentive. An intensive and expensive programme of detailed product improvement will be needed.

Some other systems - such as thermal protection, windscreens, windows, and lightweight propellant tanks - will also need a programme of life extension and maintenance reduction, but the timescales and costs are likely to be less than for rocket motors.

It therefore seems probable that spaceplanes will be approaching airliner maturity some ten years after an intensive development programme starts. The gains in maturity from sub-orbital spaceplanes will be readily transferable to later orbital vehicles.

CHAPTER 11 - MARKET

In developing the business case for spaceplanes, an important aspect is the commercial demand. Governments have shown no sign so far of being willing to fund their development, and private investors will want a high return on investment. This in turn requires high sales.

The first application of spaceplanes will be to replace existing expendable vehicles used for launching satellites and for government manned space missions. No expendable vehicle can compete in cost and reliability with a fully reusable one that can do the same job. The size of these early markets is therefore important for the spaceplane business case.

The number of satellite launches has not changed greatly over the past twenty years, at between about 70 and 130 per year. The year 2000 was typical, with 85 launch attempts, four of which were failures. Seven of these launches, all successful, were manned - five using the U.S. Space Shuttle and two the Russian Soyuz. Fifty-three of the launch attempts were for government missions and 32 were commercial, mainly for communication satellites. Twenty-two types of launcher were used, and the total cost of the launch services was around $4 billion for satellites and $3 billion for manned space missions, making a total annual market for launch services of around $7 billion.

These statistics indicate how much less mature launchers are than aeroplanes. There would not be very much flying if there were four crashes for every 85 take-off attempts. A successful type of airliner design will have more than 1000 examples in service, which between them will make more take-offs in one hour than the entire number of satellite launches in a year.

Of the 22 launcher types used in 2000, only Proton and Soyuz made a double-digit number of launches, which indicates the over-supply situation in the heavily subsidised launcher industry.

The statistics also give an indication of the problem of raising the finance for a spaceplane. Sänger development cost was estimated to be around $17 billion. If it were to capture half the present launcher market of some $7 billion per year and to make profits of 20% of sales, i.e., $700 million per year, the payback period would be some 24 years, which is far too long to attract investors. As mentioned earlier, we need to greatly reduce development cost and/or to find large new markets.

This leads to space tourism, as the least speculative large-scale commercial use of mature spaceplanes. As mentioned earlier, we know from the experiences of some 400 astronauts to date that visits to space are fascinating experiences. Thus, it seems likely that large numbers of people will want to go as soon as it is sufficiently safe and economical.

Market research pioneered in Japan [26] suggests that around one million people per year from Japan alone would pay $20,000 for a few days in a space hotel. Subsequent surveys in Canada, the United States, and the United Kingdom indicate a comparable fraction of the population prepared to pay that amount.

The Japanese Gross Domestic Product is about one fifteenth of the world total. If demand is scaled in proportion to GDP, as many as 15 million people per year might be prepared to pay $20,000 for a holiday in space. However, given the limitations of market surveys of an expensive product that requires a lot of imagination to appreciate, it would be prudent to assume an annual demand of just one million space tourists. Even this would require a fleet of around 50 spaceplanes of 50-seat capacity making one flight each per day.

As a quick check on this market estimate, one million people per year is equivalent to 7.5% of the world's industrialised population (assumed here to be one billion) making one flight per lifetime (assumed to be 75 years). This seems a conservative estimate of the likely demand for a 'once-in-a-lifetime' visit at a cost of a few months income, particularly since most people say that they would go more than once. Moreover, the fact that several people per year now wish to follow Dennis Tito and spend $20 million and give up six months of their lives to intensive training for a visit to the International Space Station is an indication that the market research is probably conservative in its estimate of demand.

One million people paying $20,000 each is an annual market of $20 billion. Adding the $7 billion market for launching satellites and for government manned space flight, the reasonably firm market for business using mature spaceplanes approaches $30 billion per year, with the prospect of far larger markets to come. This is a large business. The challenge is to produce a financially attractive development strategy.

CHAPTER 12 - DEVELOPMENT COST

This chapter explains the apparent paradox that spaceplanes can be less expensive to develop than expendable launch vehicles, despite being much safer and less expensive to fly.

When estimating the development cost of a new aeroplane, the starting point during preliminary design is usually a comparison with broadly comparable previous types. If one were designing a helicopter, for example, one would plot the development cost of previous helicopters against their empty weight, or dry mass as it is sometimes called. One would expect to find a broad trend of cost increasing with weight. Large helicopters usually cost more to develop than small ones. Placing the new design on this trend line would provide a rough preliminary estimate, which would usually be followed by a more sophisticated analysis, taking into account relative complexity and other factors. Eventually, a 'bottom up' estimate would be made, by calculating the cost of the individual development tasks and procurement items and adding them up. For present purposes, the simple method of cost versus weight will suffice.

Even at the broad-brush stage of cost estimation, there are many factors affecting development cost other than empty weight. These include the sophistication of the design, the experience of the design team, the readiness of the technology to be used, and the commercial incentive to keep costs low. A company-funded aeroplane will often cost less to develop than a government-funded one. If these factors are kept in mind, the simple cost versus weight trend lines can still provide useful insights and preliminary cost estimates.

Since operational spaceplanes are a new type of vehicle, the trend line does not yet exist and we need to look at other types of flying machine and extrapolate. Figure 12.1 shows development cost versus weight trends for six types of flying machine: high performance aeroplanes, airliners, manned spacecraft, expendable launch vehicles, demonstrator/prototype aeroplanes, and demonstrator/prototype reusable launch vehicles.

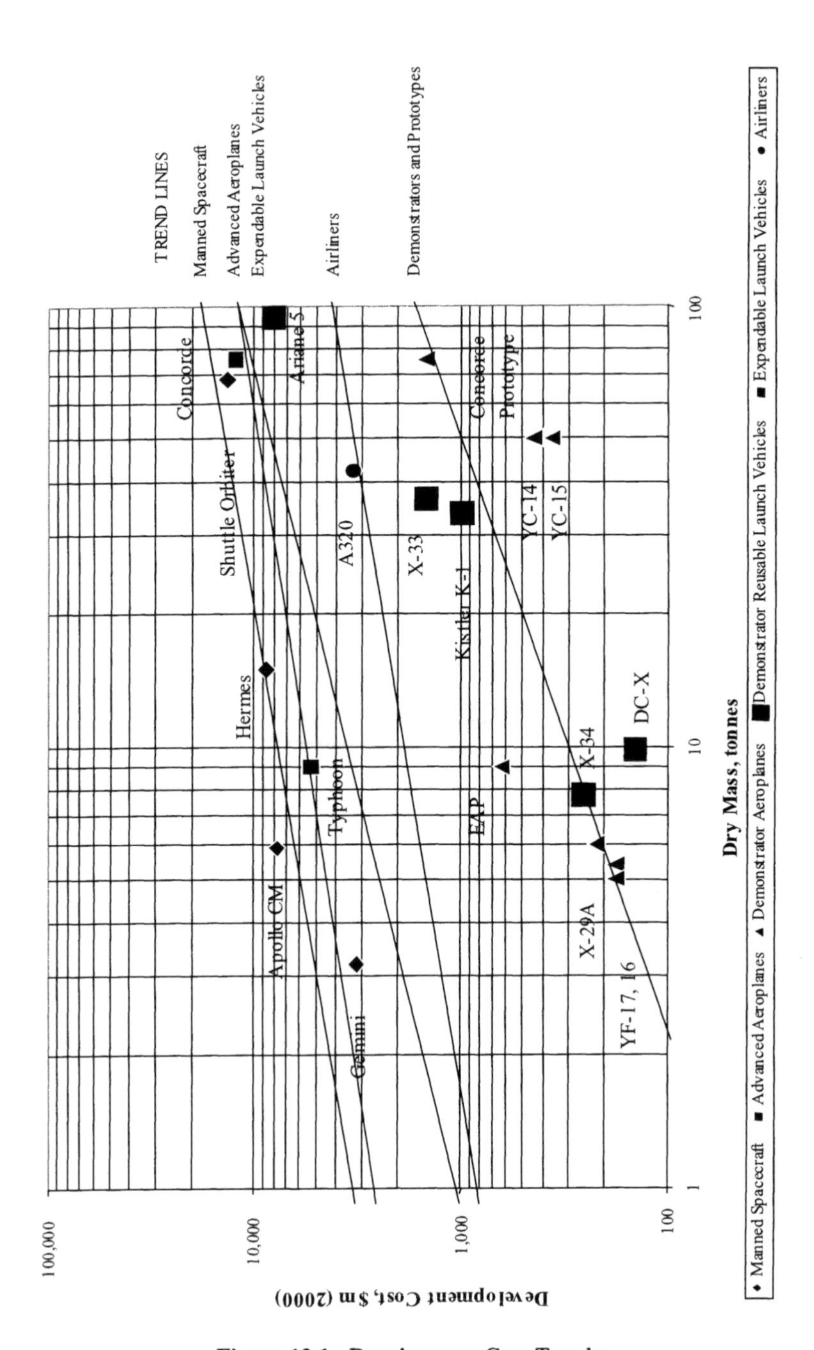

Figure 12.1 Development Cost Trends

A prototype reusable launch vehicle should be less expensive to develop than an expendable one.

Manned spacecraft are represented by Gemini, Apollo Command Module, Hermes (which never flew), and the Space Shuttle Orbiter. High performance aeroplanes are represented by Typhoon and by Concorde. Expendable launch vehicles are represented by Ariane 5. Airliners are represented by the Airbus A320. Demonstrator/prototype aeroplanes are represented by the YF-16, YF-17, X-29A, EAP, YC-14, YC-15, and the Concorde prototype. Reusable launch vehicle demonstrators and prototypes are represented by the X-34, DC-X, the Kistler K-1, and the X-33.

These aeroplanes and launchers are shown in Figure 12.2.

The trend lines for manned spacecraft, high performance aeroplanes, Expendable Launch Vehicles, and airliners are taken from a comprehensive review of development and other costs for various types of flying machine [27]. These trend lines were estimated by regression analysis using an extensive database. Even so, several of the projects shown have costs significantly different from their appropriate trend line, which illustrates the limitations of such broad-brush cost estimation.

Reference 27 does not give trend lines for demonstrator/prototypes of aeroplanes or reusable launch vehicles. That shown here has been drawn in by eye, which is sufficiently accurate for present purposes. Within the accuracy of this analysis, these two types of vehicle share a common trend line.

Before considering these trend lines, it is worth noting some of the main features of these aircraft and launchers.

Gemini was a two-seat spacecraft designed to bridge the gap between the earlier Mercury programme, in which the first U.S. astronauts went to space, and the start of the Apollo lunar landing programme. It was launched by a Titan 2. Ten missions were successfully flown in 1965 and 1966.

Gemini, Apollo CM, Hermes, Shuttle Orbiter

Typhoon, Concorde, Ariane 5, A320

DC-X, X-34, X-33, Kistler K-1

YF-16, YF-17, YC-14, YC-15

X-29, EAP, Concorde Prototype

Figure 12.2 Exemplar Vehicles Used in Figure 12.1

Prototype and demonstrator aeroplanes and spaceplanes can be built at modest cost.

The Apollo Command Module was the control centre and living quarters for the three-man crews that went to the Moon.

Hermes was a projected European mini-shuttle, designed to be launched on top of an Ariane 5. It never flew, but its development cost was well established.

The Space Shuttle Orbiter is the reusable component of the Space Shuttle. Its first orbital flight was in 1981.

Concorde is well known as the only supersonic transport in service. The only other one yet built, the Russian Tu-144, had a short and unsuccessful career. Concorde development was protracted and expensive.

The Typhoon is a collaborative fighter plane being jointly developed by several major European aerospace companies. It is due to enter full squadron service in 2006.

Ariane 5 is the latest in the Ariane series of European expendable launch vehicles. It first flew in 1996.

The Airbus A320 is a fairly recent design that first flew in 1987. Most 'new' large airliners now are in fact derivatives. Some designs still in production first flew in the 1960s, although the latest versions are much stretched, with new or improved equipment, new types of engine, and in some cases a highly modified wing.

The General Dynamics YF-16 was built to compete in a prototype fly-off competition for the USAF Light Weight Fighter (LWF) Prototype Programme. It first flew in 1974 and eventually went into production as the F-16.

The Northrop YF-17 was the other competitor in the LWF programme. It first flew in 1974. It lost to the YF-16 but was later developed into the F/A-18 Hornet for the U.S. Navy.

The Grumman X-29A was a supersonic research aeroplane intended to explore the technology of forward swept wings. It first flew in 1984.

The EAP was a technology demonstrator for the Typhoon, testing its advanced design features. It first flew in 1986.

The Boeing YC-14 was built to compete in a prototype fly-off competition for the USAF advanced medium short take-off and landing transport (AMST) competition. It first flew in 1976.

The McDonnell Douglas YC-15 was the other competitor in the AMST programme. It first flew in 1975. In the event, neither aircraft was ordered into production, but the McDonnell Douglas C-17 (now the Boeing C-17) transport uses technology developed for the YC-15.

Two Concorde prototypes were built. They were originally planned to represent the production standard and were built on production tooling more or less within the planned budget. However, weight growth and other engineering problems meant that they were unable to cross the Atlantic, and extensive stretching and re-design was needed for the eventual production standard. They first flew in 1969.

The RLV demonstrators and prototypes - X-34, DC-X, Kistler K-1, and the X-33 - were discussed in Chapter 3.

Returning to the trend lines of Figure 12.1, the most expensive vehicles to develop are manned spacecraft, followed by operational high performance aeroplanes, ELVs, airliners and demonstrator/prototypes. Conventional airliners cost about one third as much to develop as advanced aeroplanes. Demonstrator/prototype aeroplanes and RLVs cost about ten times less to develop than operational high performance aeroplanes and manned spacecraft. For example, the EAP and Concorde Prototype cost about ten times less to develop than their operational versions, Typhoon and Concorde respectively.

Why should demonstrators and prototypes cost so much less to develop than operational vehicles? The main reason is that they are not certified for operational use. They may be built to test some new technology or to improve confidence in the design or marketing. (More usually nowadays with large aeroplanes, the first of a new type is built on the production line, and will often be modified for service following the flight-test programme. Such a vehicle is not a prototype in the sense used here.)

Demonstrators and prototypes can therefore be built in experimental workshops (skunk works) using less elaborate design, testing, and tooling than is required for production aeroplanes. Moreover, they do not need low maintenance costs or long life. They are flown by test pilots only and so do not need operational standards of flight handling qualities and reliability. These factors explain why the cost of a demonstrator or prototype is about ten times less than the development cost of a fully certified operational vehicle.

That the development cost of demonstrator/prototype spaceplanes appears to share a broadly common trend line with demonstrator/prototype aeroplanes is in line with engineering intuition. Spaceplanes are much faster than even high-speed aeroplanes, but they can be simpler. They do not have to manoeuvre fiercely, fly a long distance, use short runways, carry sophisticated software for flight control, fly fast at low altitude, or have complex air intakes and nozzles for jet engines.

The development cost of a fully certified spaceplane fit for passenger carrying is likely to be somewhere between that of an advanced aeroplane and manned spacecraft of comparable empty weight. The estimated development cost of Sänger ($17 billion) is broadly in line with this suggestion. A recent Boeing study of space tourism [28] puts the development cost of the 50-seat passenger vehicle as "at least $16 billion", which is close to the estimate for Sänger.

A key point from this analysis of development costs is that a prototype spaceplane can be developed at relatively low cost - about 10% of that of an operational vehicle.

As discussed earlier, spaceplanes are potentially as safe to fly as aeroplanes, which means that a prototype spaceplane built in an experimental workshop can be piloted. Manned spaceflight to date has been more expensive than unmanned, but adding a human pilot to a prototype spaceplane should reduce development cost.

There is a fallacy in the logic that says that manned spaceflight using existing expendable launchers can be inexpensive. The argument goes that here we have a ballistic missile in service that can be stretched so that it will go to orbit, so all we need for manned spaceflight is to put a pressurised capsule on top. Such a capsule need not cost much to develop, so the programme need not be expensive.

This argument would be true if there were little requirement to recover the crew safely. Ballistic missiles are inherently unfit for human transportation because they cannot be made anything like as safe as aeroplanes. It therefore requires an enormous effort to achieve safety adequate even for a pioneering programme deemed important for national survival, or at least for national prestige. Quality control of the highest standard is needed; every system has to be designed and tested in minute detail; special crew escape systems are required in case the ballistic missile fails or blows up; and a fleet of ships is needed for retrieving the crew.

This is why the development cost of manned launch systems has been so much more expensive than those for satellites only. With a spaceplane, it will be the other way around. Having a human pilot on board will reduce development cost because it will greatly reduce or eliminate the very expensive, safety-critical, real-time software otherwise needed to control the vehicle.

A good example is provided by recent large U.S. reconnaissance aeroplanes designed for unpiloted missions, several of which have been developed in the past two decades. These vehicles have had a very high crash rate during their flight-testing programmes, mainly due to software problems. It is now common practice to add a human safety pilot for the first few flights, as shown in Figure 12.3 (although it is more usual for them to sit inside the aeroplane!). If the computer goes wrong, the pilot can switch it off and fly back to base under manual control.

Figure 12.3 Raptor Experimental Unmanned Aeroplane with Safety Pilot [Scaled Composites]
Safety pilots are sometimes used to reduce the risk, and hence the overall cost, of the early flights of unmanned aeroplanes. They can take over if the computer is set incorrectly or fails. Likewise, pilots could reduce the risk and cost of spaceplane flights.

Now imagine an unpiloted aeroplane designed to fly to and from orbit, perhaps to launch a satellite. Again, it would save money to have a human safety pilot on at least the early flights, as the cost of so doing would be less than the cost of the crashes that would otherwise be more likely. Thus, as soon as spaceplanes are developed, manned spaceflight will cost less than unmanned spaceflight.

Another thought experiment is relevant. Imagine that you manage a small offshore island that requires a daily flight by light aeroplane to deliver the mail. If the human pilot were replaced by an autopilot, you would save the pilot's salary and be able to substitute additional mail for the pilot's weight. However, with the autopilots available today, there would be far more crashes and the total cost would increase. With improved capability and reliability, autopilots will probably start to replace human pilots for such missions some day, but there is no obvious reason why spaceflight should pioneer the changeover.

Given an inherently unsafe flying machine, adding a human pilot increases the cost; given an inherently safe one, it reduces cost.

When mature spaceplanes become available, the decision whether or not to use people for a particular mission will be made on the same basis as with earthbound activities today, where there is a cost trade-off between labour and machine. The pace of automation is dictated by the advance of technology and by labour cost. By contrast, the decision to use people on space missions has so far been mainly political. The benefits may have justified the cost, at least in the early days of manned spaceflight, but the cost has been high.

There is one crucial difference between developing a prototype aeroplane and an early prototype spaceplane. Prototype aeroplanes cannot be used for revenue operations because they have not been thoroughly tested and are therefore less safe than operational aeroplanes. However, the low-cost piloted prototype spaceplane built in an experimental workshop could be used for revenue operations because it would be safer than present launch vehicles.

Before carrying fare-paying passengers, an airliner needs a type certificate, which in turn requires about 1000 test flights and full production quality control. Prototype aeroplanes do not usually meet these requirements. At present, non-certified aeroplanes are not permitted to carry even non-passenger payloads, such as cargo or cameras for air photography. This is partly due to the risks to third parties of non-certified aeroplanes and partly to maintain high safety standards in aviation. It is also due to the lack of commercial incentive for such flights, so that manufacturers have not lobbied the certification authorities for relaxation of the rules.

Prototype high performance aeroplanes are far safer and more reliable than expendable launch vehicles, and there is no reason why this should not apply to a prototype spaceplane, provided it is piloted. (One reason why both the X-33 and X-34 were cancelled was that NASA became concerned at a late stage of development by the safety implications of large, fast, unpiloted vehicles flying over the United States. Had they been piloted, the safety issue would have been far less severe. Moreover, piloted versions of both vehicles could have been converted into useful reusable lower stages of two-stage launch vehicles.)

Even a prototype piloted spaceplane should therefore be far safer and more reliable than the expendable launch vehicles that it would replace. Such replacement would therefore improve overall aviation safety and reliability, if spaceflight is counted as aviation.

Moreover, there would be a strong commercial incentive for such an 'operational prototype'. The flight-testing of the first spaceplane could be largely paid for by launching satellites or transporting professional astronauts. Such flights would be flown by test pilots under the same sort of control as airliner test flights and would be equipped with the appropriate instrumentation. Each test flight would have the objective of making a particular set of measurements, as with a normal flight test programme. While in orbit, a satellite would be released, or astronauts transferred to a space station.

In this way, the non-recurring cost to reach positive cash flow is the development cost of an operational prototype, and the spaceplane would be earning revenue when only about 10% of the total development cost had been spent. Such carriage of non-passenger payloads during flight testing would have to be approved by the airworthiness authorities, who would have good reason to do so, as they would be improving overall safety by speeding up the withdrawal of expendable launch vehicles.

Before being allowed to carry passengers, the spaceplane would have to make a large number of test flights, as does a new airliner. The carriage of non-passenger payloads would help to pay for these test flights. As spaceplane design matured and expendable launch vehicles disappeared from the scene, this concession would probably be withdrawn, and later designs would probably have to be fully certified before being allowed to carry even non-passenger payloads, as is the case with airliners today. There is therefore a strong commercial incentive to develop the first or second spaceplane. Later contenders will face a higher cost of entry into the spaceplane market because they will be competing with maturing earlier vehicles.

Thus, the development cost of a revenue-earning (operational prototype) spaceplane can be comparable to that of a demonstrator or prototype aeroplane of similar empty weight, which is less than that of an expendable launch vehicle.

The X-15 is not shown in Figure 12.1 because it was a very advanced research aeroplane rather than a demonstrator or prototype. Its development cost was about $163 million [29] which is some $1.3 billion in year 2000 money. The cost per flight in 1965, which was a typical year, was $407,000, which would be some $2.8 million in the year 2000. This high cost, equivalent to more than 20 flights of a 747, indicates the pioneering nature of the project. The X-15 was very advanced for its day. It had a nickel alloy structure designed to withstand re-entry heating, and nearly all of the equipment incorporated original features.

The X-15 had an empty weight of about six tonnes. Thus, from Figure 12.1, a prototype spaceplane of X-15 size and performance would now cost about $200 million, somewhat less than the X-34. This is equivalent to the cost of three or four jet fighters off the production line, and is perhaps the most significant conclusion from this consideration of development cost. This relatively low cost is possible because of developments over the forty-plus years since the X-15 was built.

It is worth noting that the Shuttle Orbiter, depending as it does on expendable lower stages, could not have been developed like a demonstrator aeroplane in an experimental workshop. However, the 'Sub-orbital Shuttle Aeroplane', with no expendable components, could have been so developed. The inherent lack of reliability and safety due to expendable lower stages forced the builders of the Shuttle Orbiter to use design and testing standards even more elaborate than those for operational high performance aeroplanes.

Thus, manned spacecraft on expendable launchers have the worst of both worlds. They combine the high development cost of fully certified high-performance aeroplanes with the small number of flights and technical immaturity of demonstrator aeroplanes and expendable launch vehicles.

This analysis of development cost trends has shown that the inherent safety of spaceplanes makes possible the low-cost development in an experimental workshop of a piloted prototype that could be used for launching satellites and carrying astronauts. In other words, it is precisely because spaceplanes are so much safer and less expensive to fly than expendable launch vehicles that they can cost less to develop.

CHAPTER 13 - DESIGN LOGIC

In this chapter we will derive the top-level design features and development timescales of the vehicles on the 'Aeroplane Approach' sequence discussed earlier and shown in Figure 4.1. We will start by considering the top-level design features of the 'ideal mature spaceplane', i.e., the most competitive design using technology that is theoretically achievable and on which enough research has been done to be confident that it could be made to work within a few decades. We will then work back to the preceding design and then to the one before that and so on until we derive the immediate next step. Figure 13.1 summarises these features, and the rest of this chapter shows how they were derived.

	Small Sub-orbital Spaceplane	**Small Orbital Spaceplane**	**Large Orbital Spaceplane**	**Ideal Mature Spaceplane**
No. STAGES	1	2	2	1
PILOTED	Yes	Yes	Yes	Yes
TAKE-OFF & LANDING	Horizontal	Horizontal	Horizontal	Horizontal
SEPARATION SPEED	N/A	Mach 3 to 5	Mach 3 to 5	N/A
FIRST STAGE ENGINES	Jet+Rocket (Off the Shelf)	Jet+Rocket (Off the Shelf)	Jet+Rocket (New)	Combined Air-Breathing & Rocket
PAYLOAD	2 Passengers or 250 kg	6 Passengers or 750 kg	50 Passengers or 5 tonnes	?

Figure 13.1 Basic Design Features
The top-level design features of the vehicles on the predicted development sequence can be derived using straightforward analysis.

Any prediction implies a set of assumptions. The main assumption used here is that, when the great potential of low-cost spaceplane development becomes more widely appreciated, there will be a reasonably efficient mix of government support and private sector investment, as often happens with new non-space industries. The predicted way ahead therefore assumes that the driving requirement is to achieve a mature spaceplane at low cost to the taxpayer and with a high return on investment to private investors. In other words, the predicted way ahead is what the market would choose without too much regard for the politics.

The required change in government attitude should not be underestimated. It will mean government space agencies giving up their monopoly of space policy. The Brazilian Space Agency (AEB), the British National Space Centre (BNSC), the European Space Agency (ESA), the French National Space Centre (CNES), the German Space Agency (DFVLR), the Indian Space Research Organisation (ISRO), the Italian Space Agency (ASI), the National Space Development Agency of Japan (NASDA), the Russian Aviation Space Agency (Rosaviakosmos), the U.S. National Aeronautics and Space Administration (NASA) and others will have to accept that space tourism is likely to become the largest and most popular use of space. They will have to give up detailed control of development and operations. As we will see, traditional thinking in space agencies is at present the main obstacle to the true commercialisation of space.

Nevertheless, on the assumption that a private sector culture will prevail, and with the market and engineering analyses of the preceding chapters, we can derive a reliable roadmap for the way ahead for space.

The largest business of mature spaceplanes will be serving new uses of space such as space tourism, manufacture, and solar power collection. Since these new uses are largely commercial, it is reasonable to predict that transport to and from orbit will become like an airline business, with several 'spacelines' competing for trade.

On the manufacturing front, we can expect to find a few large companies dominating the market, bringing out new or improved spaceplanes and space stations every year or two. The designs will be continually updated as dictated by the market and available technology in a manner than cannot be forecast in detail at this time. The predictions that follow are therefore limited to top-level design features.

This would be an academic exercise, perhaps like the Wright Brothers trying to predict the top-level features of the first mature airliner (the Douglas DC-3), were it not that some vitally relevant conclusions are being ignored by space agencies. Even if the Wright Brothers' predictions had been accurate (all-metal stressed-skin monoplane, radial air-cooled engines in low-drag cowlings, variable pitch propellers, flaps, retractable landing gear, cruising speed around 300 km/hour), there is not much that could have been done to speed up the development process. There would have been no way of shortcutting the years of painstaking effort needed to develop the various technologies, except perhaps that that some dead-end projects could have been avoided.

This is not the case with spaceplanes today, where the pacing factor is not technology but entrenched habits of thought.

The top-level design requirements of the 'ideal mature spaceplane' can be stated simply. Safety, reliability, turn-around time, life, and ease of maintenance should be as good as, or at least approaching, airliner standards. A vehicle meeting these requirements offers the prospect of visits to space affordable by middle-income people, and of very large new space markets.

Given that these requirements are satisfied, the factor determining the most successful design will be the cost per unit payload to orbit. In a competitive environment, the most economical design will succeed. This is analogous to the development of airliners. The first airliner that could make a profit without government subsidy was the Douglas DC-3, which first flew in 1935. Since then, operating costs have driven airliner design.

Since the X-15 demonstrated the basic technology of a spaceplane in the 1960s, there have been numerous (probably more than 100) published designs. A great variety of configurations has been proposed, the key differences being:

- One or two stages
- Piloted or unpiloted
- Rocket and/or air-breathing engines
- Vertical or horizontal take-off and landing
- If two stages, separation at subsonic, supersonic or hypersonic speed

We will now consider each of these fundamental design trade-offs in turn, from the points of view of safety and economy.

One or Two Stages

There is little doubt that eventually the most economical spaceplanes will have only one stage. Two-stagers like Sänger need two crews and, roughly speaking, have twice as many parts to fail and to be maintained. A special facility will be needed to assemble the two stages before take-off. Achieving adequate safety and reliability for the separation manoeuvre will require much design time and testing. Twice as many landing slots will be needed, at possibly congested airports.

Piloted or Unpiloted

The analysis of development cost showed that spaceplanes with human pilots would cost less to develop than those without, at least with autopilots at their present state of development. However, the ideal mature spaceplane is several decades into the future, and cockpit automation is advancing rapidly. The latest airliners are 'flown' by computer. The pilot's movements of a control stick are fed into the computer, which decides how to interpret them and sends the appropriate signals to the flying controls.

There is fascinating debate on how much authority the pilot should have to override the computer. Ancient prejudices about human autonomy conflict with rational safety analysis. Different authorities have conflicting rational safety analyses. The problems start when the pilot either does not agree with the computer or, more frequently, does not understand what it is trying to do. For all the debate, cockpit automation has added much to airline safety, and can be expected to continue so to do.

On the military side, large unpiloted and autonomous reconnaissance aeroplanes are now in service, and there are major research and development programmes aimed at unpiloted combat aeroplanes. 'Autonomous' in this context means that once launched the vehicle is entirely under its own control, following the instructions of its on-board computer. Until recently, unpiloted vehicles were either controlled by a remote pilot using a radio link or, if autonomous like missiles, expendable. The combination of reusability with autonomy is new.

In spite of such rapidly increasing cockpit automation, it is expected that airliners and business jets will require human pilots for the plannable future. Unpiloted aeroplanes to date have fallen far short of airliner safety standards. It therefore seems prudent to assume that mature spaceplanes will also need human pilots.

Rocket or Air-Breathing Engines

In the plannable future, rocket engines will be needed for the high-speed segment of an ascent to orbit. On the other hand, if all-rocket propulsion is used, the propellant fraction for a single-stage vehicle becomes impracticably high, as discussed earlier. The answer is to use air breathing for the early part of the ascent, followed by rocket propulsion for the later part. These two modes will probably be combined in one engine.

Vertical or Horizontal Take-Off and Landing

The use of air-breathing engines leads naturally to horizontal take-off, which in turn means having wings, which in turn leads naturally to horizontal landing. It would be theoretically possible to delete the wings and take off and land vertically using air-breathing engines, perhaps augmented by rockets, but this would require several times more thrust and the engines would therefore be far heavier. (The thrust of airliner jet engines at take-off is usually around one third of the maximum weight. For vertical take-off, the thrust has to be greater than the weight.)

There have been several proposals for single-stage rocket-powered reusable launch vehicles that take off and land vertically, using technology pioneered by the DC-X. Rocket engines are very close to their theoretical performance limits, and a propellant fraction significantly lower than the 87% discussed earlier is unlikely. Even so, this type of vehicle may become feasible. However, when air-breathing engines become sufficiently advanced, a horizontal take-off single-stage spaceplane with a far lower propellant fraction will also be feasible. This design will be more robust and will be able to carry a greater payload than the single-stage rocket-powered vehicle. It will therefore have a lower cost per unit payload to orbit.

Moreover, the safety challenges with vertical landing using rockets are severe. Engine failure leads to potential loss of control and well as loss of lift. Massive redundancy in engines, flight control systems, and propellant supply would be needed, probably backed up by emergency parachutes. The resulting certification cost is likely to be far greater than for a spaceplane that lands using wings to counter gravity.

Experience with aeroplanes has shown that the only justification for vertical take-off and landing is a reduction in the size of the required airfield or aircraft carrier, and this is not a driving requirement for spaceflight. Helicopters and vertical take-off jet fighters are more expensive to operate and have significantly higher accident rates than comparable horizontal take-off aeroplanes.

However, there is a natural size limit to horizontal take-off spaceplanes and an elegant simplicity to the vertical take-off, vertical landing concept. The latter may therefore find a role as a 'super heavy lift vehicle' for launching very heavy payloads that cannot readily be sub-divided for carriage in smaller vehicles.

From this analysis, we can be reasonably sure that the ideal mature spaceplane will have the following top-level design features:

- Single stage
- Piloted
- Combined rocket and air-breathing engines
- Horizontal take off and landing

A concept sketch of such a vehicle is shown in Figure 13.2. This is by no means to scale or representative of a real design and is intended only to illustrate these key design features.

Interim Projects

Having derived the top-level design features of the ideal mature spaceplane, we will now work backwards to derive the corresponding features of the interim projects shown in Figure 13.1.

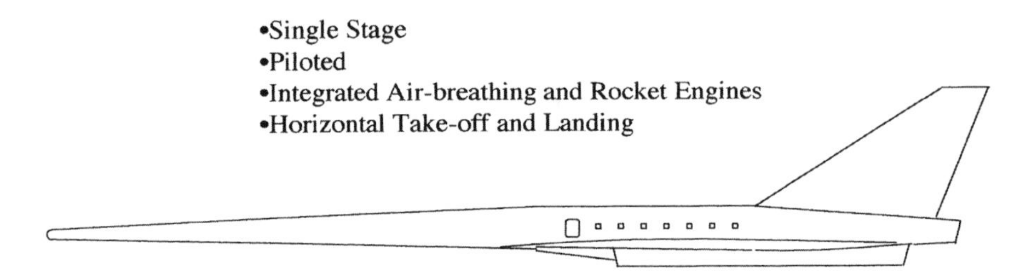

Figure 13.2 Concept Sketch of Ideal Mature Spaceplane

Large Orbital Spaceplane

Considering the top-level features of the ideal mature spaceplane, the only feature that can be relaxed to reduce development cost and time is the requirement to have a single stage. Allowing two stages greatly reduces the propellant fraction on each stage and brings the design within the scope of existing technology. The first mature spaceplane is therefore likely to have two stages.

As mentioned earlier, having two stages roughly doubles the complexity of design, manufacture and operation but, if this means bringing about low-cost access to orbit five or ten years sooner than would be possible with the single-stage ideal mature spaceplane, the market would probably decide that this was a price worth paying. Single stagers may be able to achieve a factor of two cost reduction compared with two stagers, but this is small compared with the factor of 1000 that two stagers can achieve compared with present expendable launchers, soon and at low development cost and risk. This is a case where a bird in the hand is worth two in the bush.

Having more than one stage is one of the main causes of the lack of reliability of expendable launch vehicles. Separation mechanisms and the ignition of upper stage engines are major sources of failure. However, these failures have been due mainly to immaturity, resulting from the low number of flights, due in turn to the high cost resulting from expendability.

By analogy with aviation, adequate separation safety and reliability will be achieved with spaceplanes when there is a strong enough incentive to do so. If the experimental two-stage aeroplanes mentioned earlier had entered service on early trans-Atlantic mail services, they would probably have reached safety and reliability acceptable for passenger carrying. Firing missiles from aeroplanes, dropping expendable fuel tanks, and in-flight refuelling are analogous military operations that are safe and reliable.

Considering the remaining fundamental design features, the reasons for having human pilots and horizontal landing still apply, but the use of air-breathing engines to reduce propellant fraction is no longer a fundamental requirement. Even with rockets on both stages, the propellant fraction is manageable. Thus, the main reason for horizontal take-off no longer applies.

Separation Speed

Before considering propulsion further, it is worth discussing the speed at which the upper stage separates from the carrier aeroplane - the so-called separation speed. This is an important design feature of a two-stage vehicle, and can be subsonic, supersonic, or hypersonic.

Subsonic carrier aeroplanes based on existing large transport aircraft, such as the Antonov 225 and the Boeing 747, have been proposed. The upper stage then has to provide most of the velocity increment, and requires either advanced engines or advanced materials. It is somewhat more feasible than a single-stage-to-orbit vehicle. As mentioned earlier, the Orbital Sciences Pegasus Launcher is released from a subsonic airliner. This greatly improves the flexibility of the launch site location, but does not greatly improve performance. At the other extreme, separating at hypersonic speeds up to Mach 12 has been proposed, which would involve advanced technology on the lower stage. Even the Sänger separation speed of Mach 7 requires new engines of advanced technology.

Several design studies [30, 31, 32, 33] have shown that both stages can use existing technology if the separation speed is in the high supersonic range, i.e., between Mach 3 and Mach 5. The lower stage can then be of robust aeroplane-like design.

The two fastest exemplar aeroplanes are the rocket-powered North American X-15, which reached Mach 6.33 (Mach 6.70 with a later drop tank), and the jet-powered Lockheed SR-71, which cruised at Mach 3.3. If fitted with some thermal protection and a rocket engine for acceleration to speeds above Mach 3.3, a re-designed SR-71 could reach about Mach 5 with a useful payload by using most of its cruise fuel for acceleration.

Thus, the separation speed should be the maximum achievable by a robust aeroplane with existing engine and materials technology, which is in the high supersonic region.

Returning to the choice between rockets and air-breathing engines, the upper stage should clearly have rockets only, as the main benefit from air breathing is during the early ascent. The carrier aeroplane can therefore have all-rocket, all-air-breathing, or combined propulsion. As mentioned earlier, the Sänger carrier aeroplane has all-air-breathing engines.

Jet engines are more suitable than rockets for taxiing, fly-back, aborted landings, ferry flights, and cruising to the required orbit plane; and there is a strong case for using jet engines for these phases of flight, as well as for the early boost phase. With existing technology, a maximum jet speed of around Mach 4 should be achievable, either using a turboramjet engine or a conventional jet engine with a massive injection of water or liquid oxygen into the air as it enters the engine. This would cool the air flow and increase the mass passing through the engine, thereby enabling the engine to operate faster and higher. Either way, a new engine will be required for optimum performance.

Separation speed and height can be increased by using rocket engines in addition to the jets during the boost phase of the ascent. Either these rockets can be added to the carrier aeroplane, or the upper-stage rocket engines can be used before separation, using propellant transferred from the carrier aeroplane.

Rocket engines can be used up to any altitude, and a jet-plus-rocket boost phase could therefore achieve a separation altitude higher than an all-jet phase. A rocket-powered zoom-climb would even enable separation clear of the effective atmosphere. Such a 'ski-jump' separation reduces the dynamic air loads, which otherwise present a severe design case. Separation was one of the high-risk areas on Sänger.

A further advantage of jet propulsion is that it paves the way for the ideal mature spaceplane, which uses air-breathing engines. A two-stage spaceplane with jet engines on the carrier aeroplane would be an ideal test-bed for more advanced air-breathing engines. As these engines were developed for higher speed, the maximum speed of the lower stage could be increased, leading to less propellant and higher payload in the upper stage. Eventually, the technology would enable the development of the single-stage vehicle.

Thus, the most appropriate propulsion solution for the boost phase of the large orbital spaceplane is to use new jet engines of existing technology for acceleration up to their maximum speed, augmented by massive liquid injection and/or by rocket engines to increase separation speed and height.

The use of such air-breathing engines in turn leads naturally to horizontal take-off, as with the single stage ideal mature spaceplane.

Using existing technology and with a take-off weight comparable to that of the largest jet airliner, the payload capacity is in the region of 5 tonnes. This could be 50 passengers or a medium satellite. A higher payload could be carried by using more advanced technology, but at the expense of development cost and time. The market will probably choose to use conservative technology for early large orbital spaceplanes, to be followed by progressive advance.

Summarising the above, the large orbital spaceplane is likely to have the following top-level design features:

1. Two stages
2. Piloted
3. Horizontal take-off and landing
4. Separation speed in the high supersonic region
5. New jet engines of existing technology during the boost phase, aided by massive liquid injection and/or rockets
6. Payload of about 5 tonnes

The Bristol Spaceplanes Limited Spacebus, shown in Figure 13.3, is an example of such a vehicle. It is like a Sänger designed for lower development cost and a much larger market. It uses kerosene fuel on the lower stage which, as discussed in Chapter 5, is far less expensive than the hydrogen used in Sänger. The lower stage also incorporates rocket motors to enable separation at higher speed and height, thereby avoiding the need for advanced hypersonic engines. High-altitude separation reduces the peak air-loads on the Orbiter stage, which can thereby be built with a lighter structure and carry more passengers.

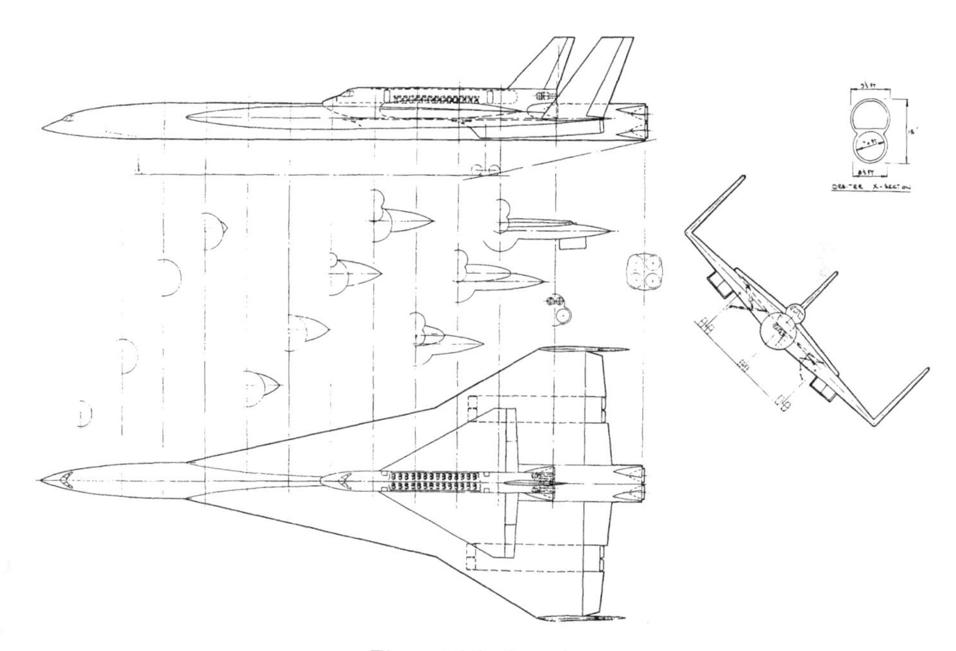

Figure 13.3 Spacebus
Spacebus is a 50-seat spaceplane designed to provide low-cost passenger transport to orbit as soon as practicable.

Spacebus has an estimated cost per seat to orbit of about $5,000 [19], which is lower than that of Sänger because it uses less hydrogen fuel and carries 50 passengers compared with 36.

Because it uses existing technology, a prototype of the large orbital spaceplane could be flying in about five years, given adequate funding. With a payload of five tonnes to low orbit, such a prototype could launch a high proportion of present satellites, using an upper stage for geostationary orbits and with assembly in orbit of payloads from several flights available for larger satellites.

It would, however, fall far short of the maturity needed for carrying passengers. Rocket engine life would be short, as would that of the thermal protection system. Moreover, it would be difficult to raise the development funding for such a large vehicle at the present stage of the market. This suggests the need for a small orbital spaceplane to develop the market and provide a focus for maturing the technology.

Small Orbital Spaceplane

Working back from the large orbital spaceplane, the two features that can be relaxed to reduce development cost and time still further are propulsion and payload size.

Propulsion

To minimise development cost, the small orbital spaceplane can use off-the-shelf jet engines. These are limited to about Mach 2.5, or one tenth of orbital velocity. (The Mach 3+ engines of the SR-71 are no longer in production.) It may be possible to increase this speed without expensive development by massive liquid injection, but the scope is more limited for an existing engine than for the optimised new one for the large orbital spaceplane. Rocket propulsion during the boost phase would allow adequate separation speed and height.

Payload Capacity

Given that space tourism is likely to be the largest market for spaceplanes and that this market is undeveloped, a favourable strategy would be to start with a small spaceplane in order to reduce risk. The first company to build a small orbital spaceplane would be in a good position to develop larger ones as and when the market demanded.

A weight of around one tonne would be adequate for most of the payloads for which an early spaceplane would be suitable, i.e., small satellites in low orbit, space station crew, space station supplies, and pioneering orbital space tourism. The resulting vehicle would have the capacity of a medium van or pick-up truck. The small orbital spaceplane is therefore likely to have a payload of around one tonne.

Summarising the above, the small orbital spaceplane is likely to have the following top-level design features:

1. Two stages
2. Piloted
3. Horizontal take-off and landing
4. Separation speed in the high supersonic region
5. Off-the-shelf jet engines during the boost phase, aided by massive liquid injection and/or rockets
6. Payload of around one tonne

As a check on the logic of working back from the ideal mature spaceplane, we can start from today and derive the top-level design features of the first successful orbital spaceplane. Previous spaceplane projects have usually been designed for minimum take-off weight for a given payload, or lowest cost per unit payload, or maximum return on investment as a stand-alone project, or technical elegance. These criteria assume more or less static technology and markets, and give incomplete answers when a dynamic change in both technology and markets is possible, leading to a thousandfold cost reduction in a decade or two.

Since most of the funding will probably have to come from the private sector, we need to find the spaceplane development strategy that leads to the ideal mature spaceplane with maximum return on investment. The first successful spaceplane should therefore be designed for minimum development cost and maximum sales while serving as a stepping-stone to the ideal mature spaceplane. The ideal long-term solution is relevant, because of the need to avoid dead-end developments. Thus, we want to retain as many of the 'ideal' features for the first spaceplane as is realistic in a shorter timescale.

Since the new markets involve carrying large numbers of people to orbit, the first spaceplane should be designed with provision for certification for carrying passengers. Without such certification, the large new markets will not be accessible. The relevant criterion to use for selecting design features is therefore early certification with minimum development cost and risk, which implies the maximum use of existing technology. The vehicle that best meets this criterion should have the best return on investment.

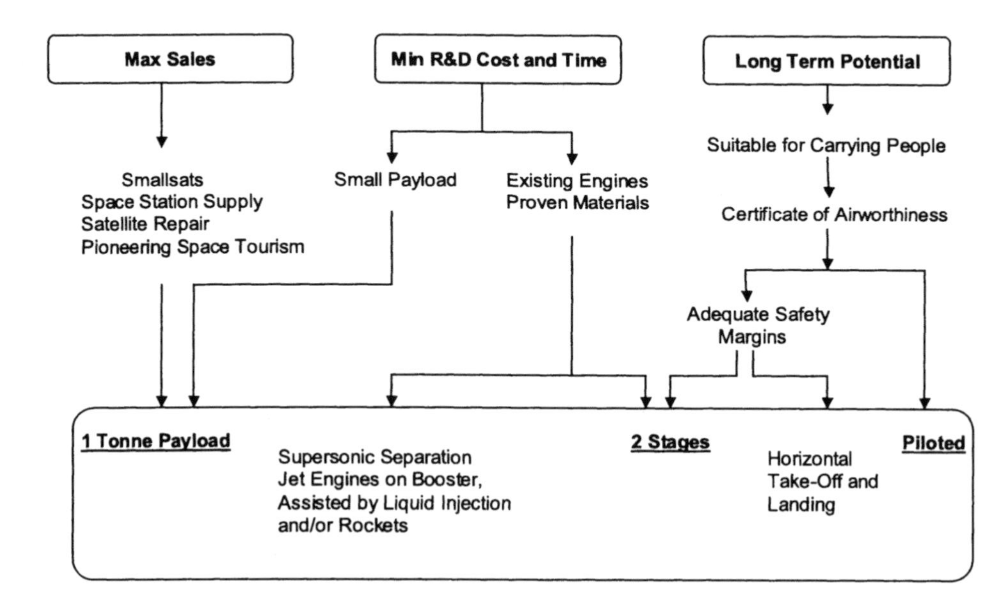

Figure 13.4 Design Logic for First Successful Orbital Spaceplane
This design logic is for early certification for carrying passengers, at low development cost and risk.

Figure 13.4 shows the logic to derive the design features to meet this criterion, which turn out to be the same as those derived by working back from the ideal mature spaceplane.

This approach adds up to making the first spaceplane as much like a conventional aeroplane as practicable, which is why this way ahead is called 'The Aeroplane Approach'. It is therefore surprising that, of the large number of recent reusable launch vehicle projects, I found just two with all these design features. Even more surprising is that in the 1960s the majority of European and U.S. projects had most of these features. Valuable insights seem to have been forgotten.

The two projects that do have all the key features are the PanAero Inc X Van [32] and the Bristol Spaceplanes Limited Spacecab [33]. As might be expected, these two projects have much in common, and it would require a detailed design study to analyse the differences and select the better features. The following notes refer to Spacecab because, as its designer, I know more about it. I am certainly not claiming here that it is a better project than X Van.

Spacecab is shown in Figure 13.5. It is in effect an update of the 1960s European Aerospace Transporter designs, using technology developed since then for other projects. The six top-level design features listed previously can be noted. The carrier aeroplane is of comparable size and shape to Concorde, but it is far simpler in detail. It only has to accelerate to maximum speed, release the upper stage, and fly back to base. It does not have to cross the Atlantic with a full payload. It does not need the last percentage point of engineering efficiency. In this respect, it is like a drag racing car that can win races with relatively crude engineering, compared with a Formula 1 or Indianapolis 500 winner that needs the highest standard of engineering sophistication.

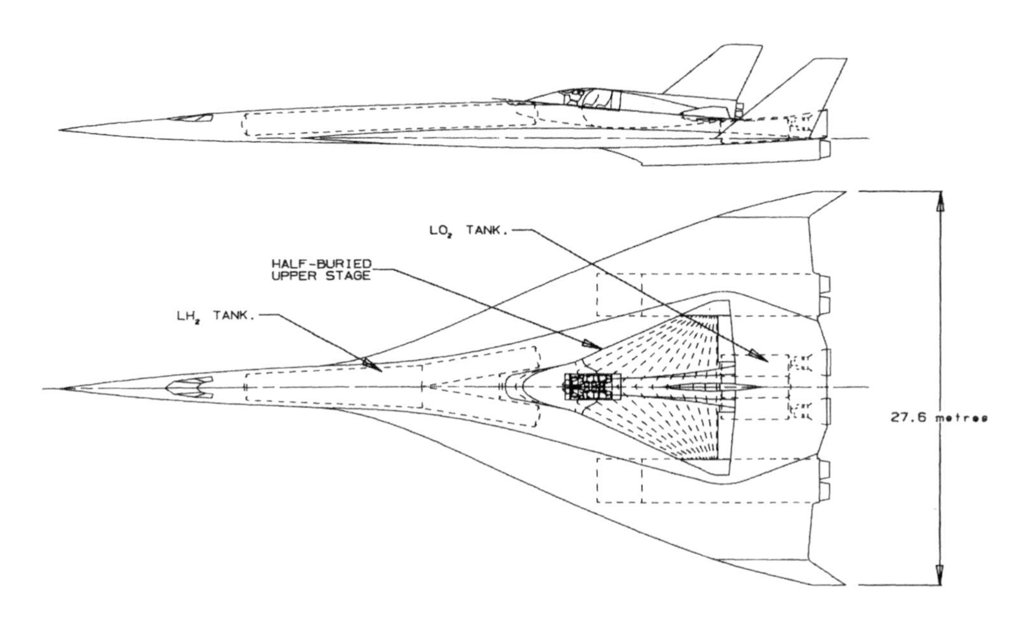

Figure 13.5 The Spacecab Spaceplane
Spacecab is designed to use existing technology.

Jet engines take Spacecab to Mach 2 and rocket engines from that speed to the separation speed of Mach 4. The rockets are used to pull up into a steep climb, from which the vehicle can zoom to a height at which the air is of such low density that air loads during separation are small. The Orbiter stage separates and then accelerates and climbs to orbit using rocket engines, while the Carrier Aeroplane flies back to base.

The payload on early Spacecab missions can be a 750 kg satellite, or six space station crew, or space station supplies. These early missions would combine flight-testing with payload delivery. After sufficient flights had taken place to demonstrate safety, Spacecab would obtain its type certificate for carrying passengers, and would then pioneer orbital space tourism.

Spacecab uses existing engines and proven materials. The development cost of Spacecab to the point of early operations is $2 billion [33]. This estimate used an earlier version of Figure 12.1.

Spacecab has much in common with Sänger, but it has about half the take-off weight and one tenth the payload capacity. At comparable levels of maturity, Spacecab would therefore have a higher cost per unit payload to orbit than Sänger. However, as discussed earlier, Spacecab is designed less for high efficiency than as a low-cost stepping-stone to the ideal mature spaceplane. Spacecab avoids the need for the advanced new engines and difficult separation dynamics of Sänger by adding rocket engines to the carrier aeroplane. Its development cost is about one tenth that of Sänger, and it has lower risk. It would be far easier to gain the development funding for an advanced vehicle like Sänger after an interim vehicle like Spacecab had demonstrated the feasibility of spaceplanes and started to build up the market.

The $2 billion development cost of Spacecab is equivalent to about two Shuttle flights, and its cost per flight would be about 100 times less on early flights and 1000 times less when mature. A Spacecab prototype could be used as a substitute for the Shuttle for supply flights to the International Space Station, several per year of which are planned over the next decade or so. Government space agencies would therefore recover its development cost in about a year.

In spite of the benefits of this 'best buy', there is no sign of space agencies showing interest in low-cost spaceplanes, and the development cost of $2 billion is probably too high for the private sector, given the political nature of present markets. There is therefore a need for an 'interim interim interim' project of even lower development cost.

Small Sub-orbital Spaceplane

The obvious such project is a small sub-orbital spaceplane. As mentioned earlier, a sub-orbital vehicle is far smaller than an orbital one for an equivalent payload, does not have to fly as fast, is in space for far less time, and has a far lower re-entry speed. Moreover, the propellant fraction required is such that a single-stage vehicle is entirely feasible. A sub-orbital vehicle is therefore far less expensive to develop.

To pave the way for a mature orbital spaceplane, the sub-orbital one should share as many top-level design features as practicable. Considering the basic features derived earlier in this chapter, there is no need to have two stages, and the smallest useful payload is one or two passengers or equivalent scientific experiment. Because there is only one stage, separation speed becomes irrelevant. There are still strong practical reasons for having jet engines for the low-speed portions of the flight, but massive liquid injection and/or rocket engines are needed to achieve a zoom-climb to space height.

The five top-level features now become:

1. Single stage

2. Piloted

3. Horizontal take-off and landing

4. Jet engines, augmented by massive liquid injection and/or by rockets

5. Payload of one or two passengers, or equivalent experiment

There are now several sub-orbital spaceplane projects in search of funding. Most of these are competing for the X-Prize [13] mentioned earlier. At the time of writing, there are 20 registered competitors. The variety of designs is as wide as that of proposals over the years for orbital reusable launch vehicles. Some have two stages, others one; some use solid rockets, others liquid; some use jet and rocket engines, others rockets only; some have vertical take-off, others horizontal. One is even launched from a balloon.

Of the five top-level design features derived here for a sub-orbital spaceplane, two are required by the X-Prize rules (that the vehicle should be piloted and that it should have a payload of two passengers). This is not the place to discuss the merits of each design, except to mention that the following vehicles have the three remaining design features (unassisted single stage, horizontal take-off and landing, jet-plus-rocket engines):

- Ascender, by Bristol Spaceplanes Limited
- Cosmos Mariner, by Lone Star Space Access Corporation
- XVan2001, by Pan Aero Inc

The Bristol Spaceplanes Limited Ascender is shown in Figure 13.6.

As with Spacecab, Ascender is described here because I know most about it but, again, I am certainly not claiming here that it is the best design. Ascender can be thought of roughly as a third-scale 'Sub-orbital Shuttle Aeroplane', weighing some twenty times less. It combines the size of a small business jet with the shape of the Shuttle Orbiter. It carries a crew of two and either two passengers or a scientific payload. It takes off and climbs to a height of 8 km using its jet engines. The pilot then starts the rocket engine and pulls up into a steep climb. At a height of 64 km the rocket propellant is used up and Ascender is climbing close to the vertical at a speed of Mach 2.8. It then coasts unpowered to a peak height of 100 km. Gravity then pulls it back towards Earth. Ascender reenters the atmosphere, pulls out of the dive, and flies back to the airfield from which it took off some thirty minutes previously.

Passengers would feel weightless for about two minutes. They would see the curvature of the Earth clearly, and an area several hundred kilometres across at one time. They would see the sky go dark with bright stars even in daytime. They would know that they had been to space.

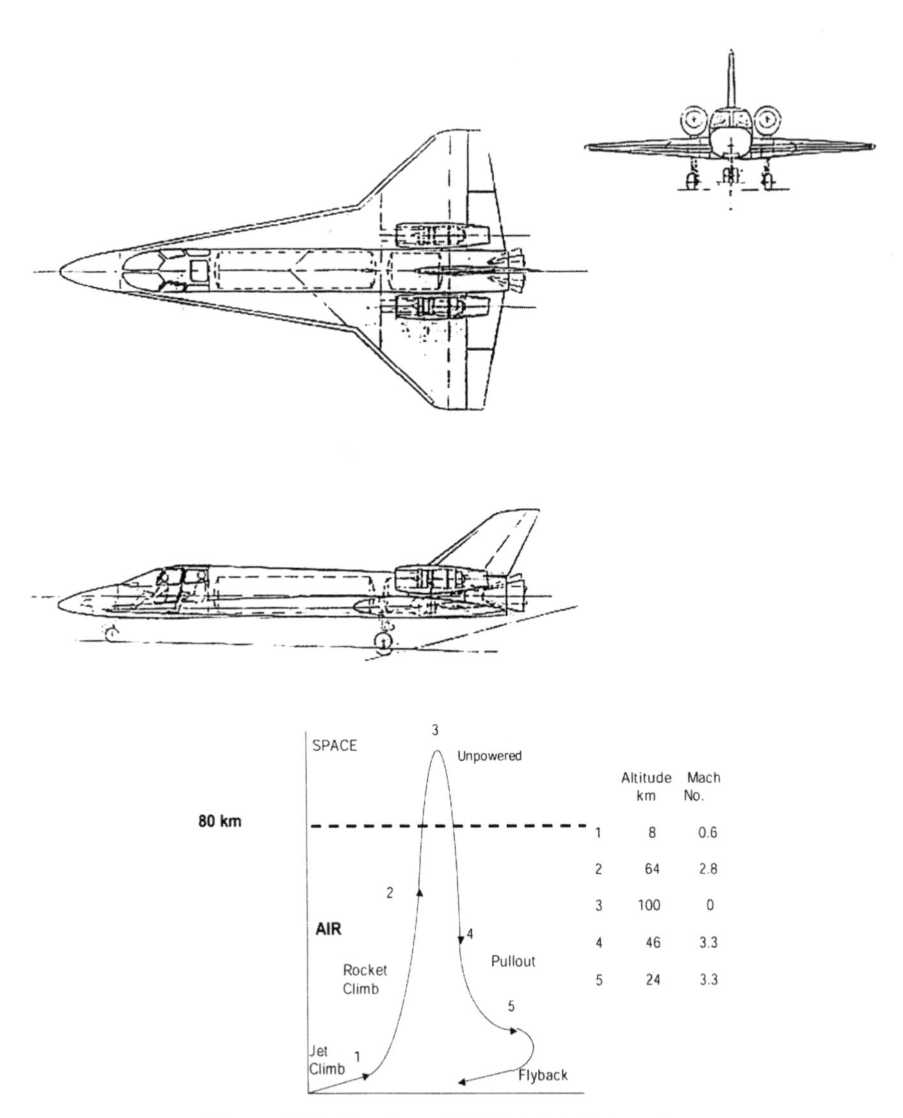

Figure 13.6 The Ascender Sub-orbital Spaceplane
A small sub-orbital spaceplane could be flying in four years for $200 million, and carrying passengers on space experience flights three years later. It is in effect an operational development of the 1960s X-15.

A technology audit of a small sub-orbital spaceplane like Ascender shows that no new technology is required for a prototype, but that considerable improvement is needed for a mature vehicle.

The shape can be scaled down from that of the Shuttle Orbiter, which is aerodynamically stable and controllable over a speed range from orbital to landing. Such a configuration has high drag at subsonic speed, but this is not critical for a spaceplane because it does not have to cruise for long distances at low speed.

The thermal protection system can again be based on that of the Shuttle, with external insulation protecting a conventional aluminium alloy structure. The Shuttle re-enters the atmosphere at about Mach 25 and takes some 20 minutes to slow down to subsonic speed. Ascender re-enters at Mach 3.3 and takes less than two minutes to slow down. The heat loads on Ascender are therefore far smaller, and more advanced insulation materials are now available.

Existing jet engines are suitable for a prototype, with some modifications to prevent the lubricating oil from vaporising in space, to protect the engines from re-entry heating, and maybe to improve re-light reliability. Existing rocket engines are also suitable for a prototype but will need development for improved safety, longer life, and lower maintenance cost before routine passenger flights can start.

A sub-orbital spaceplane requires a propellant fraction in the region of 50%, which is well within the scope of conventional structures.

Available small rocket thrusters are suitable for the reaction controls needed outside the atmosphere where aerodynamic controls are ineffective. Existing thrusters use toxic or otherwise hazardous propellants, but safer and more user-friendly propellants are under development.

The remaining systems such as navigation equipment, instruments, autopilot, communication equipment, electrical power, flying controls, and landing gear can be modified from existing mature aircraft equipment.

The development cost of a prototype small sub-orbital spaceplane should be comparable with that of other small experimental reusable launch vehicles, such as the DC-X and the X-34, i.e., in the region of $200 million, as shown earlier in Figure 12.1. Such a prototype could be used for carrying small space science experiments, cameras for high-level photography, microgravity experiments, meteorological instruments, and upper stages for launching very small satellites. It could also be used to test various spaceplane technologies, such as thermal insulation panels and advanced air-breathing engines.

After sufficient flights to demonstrate safety, a small sub-orbital spaceplane could be certified for carrying passengers. A recent market survey [34] suggests that, in the United States alone, 10,000 people per year would pay $100,000 for a flight in a sub-orbital spaceplane, making a potential annual market of $1 billion. (The Japanese market research mentioned earlier was for orbital tourism, and did not include sub-orbital flights). This should lead to a high return on investment for the first successful spaceplane entrepreneur.

Space tourism appears to be the first commercial use of space with the large potential market needed to justify the funding needed to achieve design maturity. Low-cost access to space in the short term therefore appears to depend on the development of a space tourism industry.

Timescale

A prototype of the small sub-orbital spaceplane could be flying in four years and carrying fare-paying passengers about three years later, following a development programme combining non-passenger payload flights, certification test flights, and technology maturing. It would build up the market for spaceplanes and demonstrate the ability of spaceplanes using existing technology to make several flights per day to space, albeit on sub-orbital trajectories only. It would provide a focus for developing longer-life rocket motors, thermal protection systems, and the operating procedures for rapid turnaround. This growing maturity would transfer naturally to the later vehicles on the development sequence. By flying several times to space in one day, it would demonstrate the potential of spaceplanes and enable entrepreneurs to break the mould of traditional thinking on space transportation.

A prototype of the small orbital spaceplane could be built in five years from now, but it is more likely to be built a year or two later, because finding the development funding probably depends on the confidence generated by the earlier sub-orbital spaceplane. Likewise, the large orbital spaceplane is likely to follow a few years later. This vehicle can provide mature airline-like operations, despite the complication of having two stages. The advanced single stage ideal mature spaceplane can then be developed when the market is ready to pay for it.

Thus, the pacing items for mature airline-like operations to orbit are market confidence and the maturity of rocket engines and other equipment, rather than advanced technology. Both these aspects would be well started by the sub-orbital spaceplane.

As mentioned earlier, it will take about ten years of intensive and expensive development to mature rocket engines and some other systems to airliner standards of long life and low maintenance cost. Thus, spaceplanes could be approaching airliner maturity ten years from now, given immediate priority. However, it will take at least four years for a sub-orbital spaceplane to fly and demonstrate the potential of spaceplanes, and mature spaceplane operations are more likely to start in about fifteen years.

The prospect of space tourism can be expected to drive the development of space hotels. Since these do not need new technology, development will probably take place in parallel with that of spaceplanes, and small hotels with limited facilities and suitable for short stays should be available by the time that the first orbital spaceplane gains its passenger-carrying type certificate. Larger hotels will then be built as the market demands.

Summarising this chapter, all vehicles on the predicted development sequence - The Aeroplane Approach - are piloted, take off and land horizontally, and use air-breathing and rocket engines. The ideal mature spaceplane has a single-stage, but the preceding orbital spaceplanes have two stages to allow the use of existing technology. The immediate next step is a small sub-orbital spaceplane. Each vehicle on the development sequence paves the way for the next one in terms of engineering, market building, financial credibility, and politics. Key design features lead to spaceplanes as much like conventional aeroplanes as practicable.

Had the first sub-orbital spaceplane, the X-15, been followed up by an operational spaceplane, as was proposed by several companies in the 1960s, a development sequence on these lines would probably have happened by now. Recapping from the chapter on spaceplane history, this is the 'X-15 Way' to the Spaceplane Space Age shown in Figure 4.2.

CHAPTER 14 - BREAKING THE MOULD

There is overwhelming evidence that the aeroplane approach to space transportation would lead to more benefits sooner and at far less cost than the present government approach. NASA alone is spending more than $5 billion per year on the Space Shuttle and International Space Station, whereas the private sector could transform space activities in about five years for some $2 billion by developing a small orbital spaceplane, which is some 8% of the NASA budget for manned spaceflight in that period. Initially, this would be used for supply missions to the International Space Station at a fraction of the cost of present Shuttle supply flights. This would transform the usefulness of the International Space Station, and pave the way for additional smaller space stations using International Space Station modules.

There is equally strong evidence that government space agencies are not interested in low-cost spaceplanes and space tourism. For example, NASA recently carried out a feasibility study of space tourism, jointly with the Space Transportation Association [35]. This study had very positive conclusions and recognised the revolutionary potential of space tourism. For example, the following three paragraphs are taken from page 21:

"In addressing, seriously, the possibility of our private sector providing space travel and tourism systems and services to the general public we should all appreciate that what is being discussed here is nothing less than a fundamental challenge to our views of, and participation in, extra-earth activities. It is not unreasonable to characterize this challenge as politically, socially and economically revolutionary.

"We now see the opportunity of opening up space to the general public - a "sea change" in our half-century sense that people in space would continue to be very few in number, would be limited to highly trained professionals who, at personal physical risk, would conduct mostly taxpayer supported scientific and technical activities there under government purview.

"Now the dream of very many of us during the Apollo era that we could someday take a trip to space for our own personal reasons is finally approaching realization."

In spite of such far-sighted thinking, not one of the recommendations for government action in this report has been carried out and space tourism is not even mentioned in briefings on the Space Launch Initiative.

By their reluctance to adopt low-cost spaceplanes and space tourism, NASA and the other space agencies have become the biggest obstacles to true space commercialisation. Large aerospace corporations tend to take their lead from government space agencies or defence ministries and, at the time of writing, none of these is actively pursuing space tourism.

Why should NASA be so slow to take seriously the prospects for low-cost spaceplanes? Useful insight on the motivation of large government agencies [36] suggests that they are driven primarily to defend their budgets and that, if they can get away with poor delivery, they will. It is as if NASA were saying to itself, "When travel to orbit becomes an airline business, who will then need NASA?" It is as if they perceive a threat to the status quo and to their budget. This is a short-term view. The probable answer to this question is that when airlines do indeed take over travel to and from orbit, there will be an even bigger need for NASA. The cost of exploring the solar system will then be slashed but, until there is profit in going there, government agencies will be needed. Space tourism will probably generate an even greater public enthusiasm for space exploration, and government space agencies should be able to increase their budgets and rapidly explore the solar system, initially by robots and then by people. There is no shortage of places to go. A lunar base and manned mission to Mars, long the dream of NASA, would become far more affordable than today.

For some three decades, NASA and the other government space agencies have in effect suppressed the development of spaceplanes and thereby of low-cost access to orbit. There is no evidence of a conspiracy as such. Rather, it is a question of how large organisations with huge reserves of prestige and patronage, and paid for by the taxpayer, react almost subconsciously to the threat of change. It is a question of internal culture.

There is a useful comparison with the development of aeronautics. Until 1783 the human race was earthbound. Then the Montgolfier Brothers built the first man-carrying balloon. Between then and 1903 the only way to take to the air was in such vehicles, which had the serious limitation of not being able to fly into wind. Balloons were useful only for specialised purposes such as artillery spotting, meteorological research, and pleasure flights.

Then the Wright Brothers flew the first aeroplane. This showed that flying machines could travel in any direction potentially at high speed, thereby opening up the prospect of large-scale military and commercial flight. This achievement led to an explosive growth in aeronautics.

Present launchers can fly once only, which is a fundamental limitation analogous to that of balloons being able to fly down wind only. Again, aeroplanes provide the solution - spacefaring ones in this case. Spaceplane pioneers predict a similar explosive growth in space travel when the potential of spaceplanes becomes more widely appreciated.

However, there are several differences between the pioneering days of aeroplanes and now. Early aeroplanes were within the scope of wealthy individuals, whereas early spaceplanes are not. The balloon industry was not dominated by large monopolistic government agencies. Substantial engineering problems had to be solved before the pioneering aeroplanes could be developed into useful machines, whereas the technology for a useful spaceplane was demonstrated forty years ago by the X-15. Thus in 1903 the main challenges facing aeroplane pioneers were technical - today the main challenges facing spaceplane pioneers are political.

What then will break the mould of thinking on space policy? The case for the aeroplane approach to low-cost spaceplane development is so overwhelming that it is almost certain to happen as soon as a critical mass of people take it seriously. From the experience of the private sector spaceplane and space tourism movement over some two decades, this will take more than rational argument, books, technical papers, displays at exhibitions, conferences, lobbying politicians and civil servants, or even small-scale flying demonstrators. It will probably require a piloted sub-orbital spaceplane to fly before the activists can persuade companies or governments to put up the required investment for orbital spaceplane development.

This 'trigger project' could either be one of the X-Prize candidates or a military spaceplane. One of the more likely projects at the time of writing is the U.S. Defence Advanced Research Projects Agency RASCAL (Responsive Access, Small Cargo, Affordable Launch).

This will be a small sub-orbital spaceplane capable of launching expendable upper stages for carrying small satellites to orbit, rather like the X-15/Blue Scout proposal of 1961. Competitive bids are in progress at the time of writing and the project appears to be funded up to the demonstrator phase. The budget is $88 million, which is further evidence that a useful sub-orbital spaceplane can be built for about $200 million.

RASCAL is the first funded demonstrator reusable launch vehicle intended for development into an operational version, fifty-seven years after the piloted winged V-2 study illustrated the basic feasibility of such a concept. If it is successful, it could become an historic project. RASCAL is scheduled to fly in 2006.

The first sub-orbital spaceplane, whether privately or government funded, would provide incontrovertible evidence of the benefits of spaceplanes. It would be capable of several flights per day to space at costs far closer to airliners than to present launchers. This would open the way for space tourism entrepreneurs to gain the investment for certifying the vehicle for passenger carrying, and sub-orbital space experience flights would then start.

Supporters of low-cost spaceplane development could then point out that the next step - the development of the small orbital spaceplane - would save billions of taxpayer dollars, compared with present government space plans, by reducing the number of Shuttle flights needed for International Space Station supply. It would provide low-cost access to International Space Station on demand, thereby greatly improving its usefulness and slashing the cost of additional space stations using International Space Station modules. Government agencies would find it difficult to resist such argument, in the face of the daily evidence provided by the sub-orbital spaceplane flights.

Once the small orbital spaceplane has flown, subsequent developments can be paced by the growth in market demand, and the critical breakthrough to low-cost transport to orbit will have been achieved.

A way ahead for low-cost spaceplanes along the general lines of the aeroplane approach shown in Figure 4.1 therefore seems likely to start soon. Historians in ten or twenty years time will no doubt be intrigued by why it took so long. They will spend many happy hours dissecting the internal culture of government agencies charged with promoting space commercialisation but in fact providing the greatest obstacle.

Perhaps surprisingly, the United Kingdom is well placed to lead the spaceflight revolution. Until the early 1970s, UK rocket technology was not far behind that of the United States and Soviet Union, although activities were on a smaller scale. Since the cancellation of Blue Streak and Black Arrow, UK policy has been to have as little to do with launchers as possible, largely on the grounds of cost. This may have been no bad thing with regard to expendable launchers, but it means that there is strong resistance even to the study of spaceplanes. However, the United Kindom has all the required technology for a useful small sub-orbital spaceplane, the closest yet to such a vehicle having been the Saunders Roe SR.53 rocket fighter of 1957. The UK government was the first to endorse a study of a low-cost spaceplane (the European Space Agency-funded study of Spacecab, mentioned earlier). The United Kingdom is the only major industrial country not involved with development of the International Space Station, having declined an invitation to take part on the grounds that it offered poor scientific and commercial return for the money. The United Kingdom has but a minor industrial involvement with expendable launch vehicles. The UK vested interests opposed to the development of low-cost spaceplanes are therefore less significant than in most other industrial countries.

In the year 2000, the Trade and Industry Select Committee of the House of Commons carried out an inquiry into UK space policy. Among the recommendations in their report [37] was that there should be a review of UK launcher policy carried out by a body independent of the British National Space Centre. This may be the first time that the space policy of any government has been so fundamentally challenged by its parliamentarians.

The United Kingdom has therefore made a good start towards a commercial space policy by not backing expensive prestige manned space projects. Whether she follows up this advantage remains to be seen.

CHAPTER 15 - BENEFITS

Any major new development is bound to cause controversy, so it is relevant to consider the disadvantages and benefits of a greatly reduced cost of access to space and of the growth of a large space tourism industry.

The main disadvantage will be increased atmospheric pollution. Large numbers of spaceplanes climbing to space will cause non-negligible injections of pollutant into the high atmosphere. If liquid hydrogen and oxygen propellants are used, the exhaust product will be water vapour, which is non-toxic. It will, however, be persistent, and the effects will need to be studied carefully.

The first benefit will be that existing space applications will become less expensive and thereby more widely available. Satellites have transformed communications, navigation, weather forecasting, remote rescue, astronomy, and Earth science: most would agree that these developments have been highly beneficial.

The second benefit is that spaceplanes should lead to a new golden age of astronomy and Earth science. There will be a rapid gain in environmental knowledge, which will help us to restrict our polluting activities in a more effective manner. Pollution caused by the spaceplanes themselves is unlikely to be near the top of the action list because of the relatively small scale of activity and the environmental benefits of low-cost access to space.

The third benefit will be new jobs and economic expansion. Space tourism in the near term and space manufacture and solar power later on will create large numbers of new skilled jobs, thereby stimulating the economies of industrialised countries and enabling them to do more to help less well developed countries.

The fourth benefit is the introduction of hydrogen fuel for air and ground transport, which will greatly reduce pollution. Hydrogen has several advantages as an aircraft fuel. It has nearly three times the heat of combustion of kerosene fuel, which greatly reduces the fuel weight needed by an aeroplane. On the other hand, it is expensive and requires very large tanks because of its low density. Hydrogen is probably inherently as safe as or even safer than kerosene, and several aeroplanes experimentally converted to burn hydrogen fuel have flown successfully. It is used in launch vehicles, including the Space Shuttle. The technology is, however, immature, and very few aircraft designers are familiar with it. Much detailed development will be needed to achieve the potential safety. Space tourism is likely to be the first large-scale commercial use of hydrogen fuel, which will help to break the familiarity barrier and bring forward its use in aeroplanes. This in turn will make it easier for hydrogen fuel to be adopted for ground transport systems.

These disadvantage and benefits can be reliably expected. The remaining benefits are more speculative because they involve human psychology. What will be the effects on human thinking of large-scale space tourism, space art, space sport, and transformed space science?

Some indication is available from the experience of astronauts, more than 400 to date. Many have said that going to space was a transforming experience and that they would like to go again, [38]. They tend to return to Earth with a more global perspective than when they left. They are more conscious of the fragility of 'Spaceship Earth'.

When a million and more people visit space each year, we can expect these views to spread, which should make it easier to generate global action to counter the various perils that face our home planet.

Many people consider that the main benefit from the Apollo lunar landing programme was the famous 'Earthrise' photograph, Figure 15.1, of the Earth appearing to rise above the Moon. To quote from a senior NASA Apollo engineer, Henry O Pohl, looking back on Apollo:

"Probably the most significant benefit of the Apollo programme was pictures from space, allowing everyone see the Earth for what it was - a little ball with a very thin, fragile atmosphere around it. One picture from Apollo of the whole Earth caused the entire world to start thinking about what we were doing to the environment and the planet. That picture may have done more for mankind than any other single thing - just giving us a perspective of the world we live in. That's something we don't normally think about. For the first time man could see the Earth all at one time and realize it as a very small, very fragile planet, and we were really destroying it fast" [39].

I well remember seeing 'Earthrise' for the first time and, like millions of others, starting to take the environmental movement seriously.

Figure 15.1 'Earthrise' [NASA]
Taken from Apollo 8 by Frank Borman in December 1968, this photograph soon became the icon of the environmental movement.

There is no shortage of articles and books about the future of the human race in space. Some suggest that space colonisation is the next breakthrough in evolution, like the first animals coming out of the sea to colonise the land. Others suggest that space provides the answer to the 'limits to growth' on the home planet, which are already leading to increased regulation, rationing, and reduced individual freedom. Others point out that a frontier, now all but gone on Earth, is needed as an outlet for restless spirits, and that massive space exploration might restore a sense of optimism to human endeavour.

Perhaps Tennyson had something like this in mind when in 1837/8 he wrote in 'Locksley Hall':

"For I dipt into the future, as far as human eye could see,

Saw the Vision of the world, and all the wonder that would be;

Saw the heavens fill with commerce, argosies of magic sails,

Pilots of the purple twilight, dropping down with costly bales;

Heard the heavens fill with shouting, and there rained a ghastly dew

From the nations' airy navies grappling in the central blue;

Far along the world-wide whisper of the south-wind rushing warm

With the standards of the peoples plunging through the thunder storm;

Till the war-drum throbbed no longer, and the battle-flags were furled

In the parliament of man, the Federation of the World."

CHAPTER 16 - CONCLUSIONS

We have considered the way ahead for space transportation from the points of view of the market, engineering, economics, history and politics. In each of these fields, we have considered a straightforward analysis. The synthesis of these analyses points strongly to an impending spaceflight revolution. While some of the argument is inevitably speculative, and some of the detail likely to change, the following conclusions appear to be soundly based:

1. An operational prototype of a fully reusable aeroplane-like launch vehicle can be built in about five years using existing technology for the cost of approximately two Space Shuttle flights.

2. The cost of developing such a prototype can be lower than that of current expendable launchers, precisely because spaceplanes are so much less risky and expensive to fly. The inherent safety of aeroplane-like flying machines allows their development like demonstrator aeroplanes in an experimental workshop. The low cost per flight permits a full aeroplane-like flight test programme.

3. Such an operational prototype could be used for supply operations to the International Space Station. It would be so much less expensive to fly than the Space Shuttle that government space agencies would recover the development cost within a year of its entry into service. Moreover, low-cost access more or less on demand would transform the usefulness of the International Space Station.

4. Enlarged and mature developments would have a cost per seat to orbit about 1000 times less than that of the Space Shuttle.

5. Costs that low could be achieved in about fifteen years. The pacing item is probably the development of a long-life rocket engine.

6. A strong commercial incentive is needed to enable this process of

enlargement and maturity. Space tourism is likely to provide this incentive and to become the biggest space business. Brief sub-orbital flights should become widely affordable within ten years, and visits to orbiting hotels within fifteen years.

7. The resulting slashing of the cost of access to orbit would benefit all present users of space. Astronomy and environmental science in particular would be transformed by very large new instruments in space.

8. There are no insurmountable problems in achieving safety approaching that of airliners.

9. This aeroplane approach to spaceplane development leads to low-cost access to orbit far sooner and at far less cost than do present government space plans.

10. Fully reusable launchers were widely considered both feasible and the logical next step in the early 1960s. Many of the designs were inspired by the very successful X-15 research aeroplane, which is still the only fully reusable vehicle to have flown to space height.

11. The reasons why reusable launchers have not yet been developed are complex, but are primarily due to the short-term vested interests and entrenched habits of thought of government space agencies, who have steadfastly managed to avoid backing promising designs over three decades. This suppression has not been due to conspiracy as such, but to the internal culture of these organisations resulting from the politics affecting their budgets.

12. Breaking the mould of present thinking and bringing about a change in policy will probably require the successful flight of a small sub-orbital

spaceplane of X-15 performance and capable of several flights per day to space. There are several promising designs for such a vehicle, and a prototype could be built for the cost of three or four jet fighters off the production line.

13. Spaceplanes will enable large numbers of ordinary people to visit space. If they react like the astronauts to date, there is likely to be increased pressure for effective action on protecting the environment and on other global challenges.

APPENDIX 1
COMPARISON BETWEEN SÄNGER AND BOEING 747

	747	Sänger Carrier Aeroplane	Sänger Orbiter	Sänger Total
Technical Data				
Wing Span, m	65	46	18	
Length, m	71	92	33	
Wing Area, m^2	525	880	135	
Passengers	420		36	36
Fuel Mass, tonnes				
Liquid Oxygen			72	72
Liquid Hydrogen		119	11	130
Kerosene	137			
Total Fuel Mass	137	119	84	203
Empty Mass	184	171	29	200
Take-off Mass, tonnes	363	290	115	405
Cost Data				
Complexity Factor	1	2.5	5	
Fuel Unit Cost, $/kg				
Liquid Oxygen	0.11			
Liquid Hydrogen	2.80			
Kerosene	0.26			
First Cost, $ million	130	302	102	
Flights per day	1	2	1	
Annual Costs, $ million				
Amortisation	13	30	10	
Insurance	2	5	2	
Costs per Flight, $				
Amortisation	35,616	41,375	28,067	69,442
Insurance	5,342	6,206	4,210	10,416
Crew	18,000	18,000	18,000	36,000
Maintenance	15,600	36,245	12,293	48,538
Fuel	35,620	333,200	38,720	371,920
Landing Fees, Navigation	10,000	10,000	10,000	20,000
Total	120,179	445,026	111,291	556,317
Cost per Seat	286			15,453

Notes

1 In the above table, complexity factor is a measure of relative production cost per unit mass. Thus the Sänger Orbiter is assumed to cost five times

as much per kg empty mass as the 747.

2 The cost per kg of liquid hydrogen assumes a higher production rate than at present. Even so, it is some ten times more expensive than kerosene.

3 First costs have been scaled by empty weight and complexity factor.

4 The 747 is assumed to make one 12-hour flight per day, as is the Sänger Orbiter. The Sänger Carrier Aeroplane has a flight time of one to two hours and is assumed to make two flights per day.

5 Annual amortisation is 10% of first cost, and annual insurance is 1.5% of first cost.

6 Crew costs are assumed the same for all vehicles. The shorter flight time of Sänger is assumed to balance higher salaries for spacefaring pilots and cabin staff.

7 Maintenance costs are scaled in proportion to first cost.

APPENDIX 2
PROPELLANT MASS FRACTION ESTIMATES

Orbital Velocity

For a satellite orbiting the Earth in a circular trajectory, the downward force of gravity exactly balances the outward centripetal force, so that the net acceleration along the line between the satellite and the centre of the Earth is zero, i.e.,

$$m \times g = m \times v^2 / r \tag{1.1}$$

where:

m = mass of satellite

g = local acceleration due to gravity

v = satellite velocity

r = radius of circular orbit

Eliminating m and rearranging:

$$v = \sqrt{r \times g} \tag{1.2}$$

At the surface of the Earth:

$$g = g_0 = 9.81 \text{ m/s}^2$$

$$r = r_0 = 6378 \text{ km}$$

and

$$v_0 = \sqrt{9.81 \times 6{,}378{,}000} = 7910 \text{ m/s} \tag{1.3}$$

Thus, if there were no atmosphere and the Earth's surface were absolutely smooth, a body projected horizontally at 7910 metres per second would become a satellite at zero height.

At greater heights, the local acceleration due to gravity is inversely proportional to the square of the distance from the Earth's centre, in line with Newton's law of gravity. At a height of h km, therefore, orbital velocity is given by:

$$(v_h)^2 = r \times g = (r_0 + h) \times g_0 \times (r_0 / (r_0 + h))^2$$

$$= \frac{g_0 \times r_0^2}{r_0 + h} = {v_0}^2 \times \frac{r_0}{r_0 + h}$$

$$v_h = v_0 \times \sqrt{\frac{r_0}{r_0 + h}} \qquad (1.4)$$

At a height of 200 km, for example

$$v_{200} = 7910 \times \sqrt{\frac{6378}{(6378 + 200)}} \qquad (1.5)$$

$$= 7789 \text{ m/s},$$

or 7.8 km/sec in round numbers, which is the value mentioned in Chapter 8.

This is an approximation, because the Earth is not a true sphere and does not have a uniform mass distribution. These variations become important when reducing the miss distance of a ballistic missile or predicting the precise orbit of a satellite, but are not important for first order sizing and cost comparison of different types of launch vehicle.

Rocket Equation

Consider a rocket-propelled vehicle of mass m kg, velocity v m/s, and rocket efflux velocity Ve accelerating in free space. If the only change in mass is due to the consumption of the rocket propellant, then the thrust T, measured in Newtons, is given by:

$$T = Ve \times \dot{m}$$

The acceleration is given by: $\dot{v} = \frac{T}{m}$

$$\therefore \quad \dot{v} = \frac{Ve \times \dot{m}}{m} \tag{1.6}$$

Thus, the acceleration increases as the mass reduces due to propellant consumption. Integrating between initial conditions $(\ )_i$ and final conditions $(\ )_f$ gives:

$$v_f - v_i = Ve \times \ln \frac{m_i}{m_f} \tag{1.7}$$

The velocity increment is sometimes called Δv, and the ratio of initial to final mass, m_i/m_f, is often called the mass ratio, R. The classic rocket equation then becomes:

$$\Delta v = Ve \times \ln R \tag{1.8}$$

The efflux velocity Ve is therefore a measure of the efficiency of a rocket motor in converting propellant consumption into velocity increment. It is sometimes more convenient to measure thrust in units of kg force rather than Newtons. The thrust needed for lift-off is then just greater than the vehicle weight in kg. The useful measure of rocket efficiency is then specific impulse, *Isp*, rather than *Ve*. *Isp* is the thrust measured in kg force divided by the mass of fuel consumed measured in kg per second, and which has units of seconds. *Isp* is equal to *Ve* divided by 9.81 m/s^2, the acceleration due to gravity at the Earth's surface. The rocket equation then becomes:

$$\Delta v = Isp \times 9.81 \times \ln R \qquad (1.9)$$

Propellant Mass Fraction

The practical application of the rocket equation is to estimate the propellant mass fraction needed for a given velocity increment. If the initial mass is m_i, the propellant mass m_{prop}, and the only change in mass is propellant consumption, then the final mass m_f is $m_i - m_{prop}$, and:

$$R = \frac{m_i}{m_f} = \frac{m_i}{m_i - m_{prop}} \qquad (1.10)$$

and the propellant fraction is given by:

$$\frac{m_{prop}}{m_i} = 1 - \frac{1}{R} \qquad (1.11)$$

As an example, consider a single stage to orbit vehicle using liquid oxygen and kerosene propellants with a specific impulse of 330 sec, which is typical for a modern motor, to launch a satellite to a 200 km orbit. The required velocity increment is the orbital velocity of 7.8 km/sec plus typically 1.5 km/sec for drag and gravity losses, making a total ideal velocity of 9.3 km/sec. The mass ratio is then given by:

$$\ln R = \frac{9300}{9.81 \text{ x } 330}, \text{ and } R = 17.7 \qquad (1.12)$$

The propellant fraction is given by:

$$\frac{m_{prop}}{m_i} = 1 - \frac{1}{17.7} = 94\% \tag{1.13}$$

Thus the mass ratio is nearly 18, and 94% of the take-off mass has to be propellant. This is the value shown in Figure 8.4 and, as mentioned in the text, is impractically high because it leaves only 6% for structure, engines, equipment, and payload.

With a two-stage vehicle, the velocity increment on each stage can be halved to 4.65 km/sec, and repeating the above calculation with this value shows that the propellant fraction on each stage is reduced to the manageable value of 76%, again as shown in Figure 8.4. These values are also used in Figure 8.5.

With hydrogen fuel, a specific impulse of 460 sec is practicable, and repeating these calculations shows that the propellant fraction in each stage is then 64%, again as shown in Figure 8.4. This is well within practical limits, and a reusable two-stage vehicle is feasible with existing technology.

Thus, the basic challenge of building a launch vehicle, the need for staging, and the feasibility of a two-stage launcher using hydrogen fuel, can all be demonstrated using very simple equations and calculations.

Analogy with Long-Range Aeroplanes

As mentioned in Chapter 8, useful insights into the problem of spaceplane design can be gained by comparison with long-range aeroplanes, because both share a requirement for a high propellant mass fraction. The aeroplane equivalent to the rocket equation is the Breguet range equation:

$$Range = Speed \times Lift \big/ Drag \times Isp \times \ln \frac{m_i}{m_f} \tag{1.14}$$

The difference is that with a launch vehicle most of the work done by the engine is used to accelerate the vehicle, whereas with an aeroplane most of the work is used to overcome air resistance (drag). The lift to drag ratio of an aeroplane in the cruising condition is largely independent of weight. The derivation of the equation is similar, except that the resisting force is drag rather than mass times acceleration. Thus, the output from the rocket equation is velocity increase, whereas that from the Breguet equation is distance flown. Both outputs increase with *Isp* and with mass ratio.

Care is needed with the units, and engine efficiency is usually expressed as specific fuel consumption, which is the reciprocal of *Isp* using hours rather than seconds as the unit of time. *Isp* has been used here to emphasise the commonality between the two equations.

The longest-range aeroplane to date, and the one with the highest fuel fraction, is the Rutan Voyager, shown in Figure 8.10, which flew round the world non-stop without refuelling in 1986. It had an advanced structure that weighed less than 10% of the take-off weight. As a result, it could carry 72% of its take-off weight as fuel. For the actual record, the fuel consumed was 70% of take-off weight.

The structural efficiency of a spaceplane with a 70% propellant fraction is therefore very roughly comparable to that of an aeroplane that can fly round the world once. We can now find the 'equivalent aeroplane range' of the single-stage and two-stage launchers using kerosene and hydrogen fuel shown in Figure 8.4, as in the following table:

Table 3

TYPE	FUEL FRACTION %	MASS RATIO	RANGE/EARTH CIRCUMFERENCE
Voyager	70.1	3.35	1.00
Single-Stage, Kerosene Fuel	94.3	17.69	2.38
Single-Stage, Hydrogen Fuel	87.3	7.85	1.71
Two-Stage, Kerosene Fuel	76.2	4.21	1.19
Two-Stage, Hydrogen Fuel	64.3	2.80	0.85

The propellant fractions for the different types of launch vehicle are the same as in Figure 8.4. The mass ratio is one divided by one minus the propellant fraction (expressed as a fraction rather than as a percentage), and the range, measured in Earth circumferences, is scaled from Voyager by the logarithm of the mass ratio. This is the basis for the table shown in Chapter 8.

REFERENCES

1 "Review of European Aerospace Transporter Studies", by H Tolle. Proceedings of SAE Space Technology Conference, Palo Alto, California, May 1967.

2 "Raketenflugtechnik", by E Sänger. Verlag von R Oldenbourg, München und Berlin, 1933.

3 "The Silver Bird Story" by Irene Sänger-Bredt. Journal of the British Interplanetary Society, London, 1971.

4 "A Vertical Empire - The History of the UK Rocket and Space Programme, 1950-1971", by C N Hill. Imperial College Press, 2001

5 "Report on A.T.O Tests Using D.H. "Sprite" Rocket Motors In D.H. "Comet" Aircraft." by R A Grimson. The de Havilland Engine Co. Ltd. 6th October 1952.

6 "Countdown - A History of Spaceflight" by T A Heppenheimer, Wiley, 1997. Page 258 et seq.

7 "ISS costs may reach $95 billion", Flight International, 30 January-5 February 2001, page 33.

8 "Economics of Rocket-Propelled Aeroplanes", by R Cornog. Aeronatical Engineering Review, September 1956.

9 "The Rocket Propelled Commercial Airliner", by W R Dornberger. University of Minnesota, Institute of Technology, Research Report No 135, November 1956.

10 "Boost Glide Vehicles for Long Range Transport", by D M Ashford. J R Ae Soc, July 1965.

11 "Economic Trade-Offs for Spaceplanes Over A Large Range Of Projected Traffic", by Jay P Penn and Charles A Lindley", The Aerospace Corporation, 2000.

12 Letter from Ian Taylor MBE MP, Parliamentary Under-Secretary of State for Trade and Technology, to the Rt Hon Sir John Cope MP, March 1995.

13 X-Prize Foundation, 5050 Oakland Avenue, St Louis, Missouri, 63110, USA, http://www.xprize.org

14 "Aviation. An Historical Survey from its Origins to the End of World War II", by Charles H Gibbs-Smith, HMSO, London, 1970

15 "Slow Road to Reusability", Flight International, 1 January 2000, page 106.

16 "Space Shuttle Value - Open to Interpretation", by Roger A Pielke Jr. Aviation Week, 26 July 1993.

17 The Sänger design evolved over the years. The data used here is taken from "Sänger - An Advanced Launcher System for Europe" by D E Koelle and H Kuczera, IAF-87-207, presented at the 38th Congress of the IAF, Brighton, UK, October 1987.

18 A good description of Sänger is given in Chapter 8 of "Spaceflight in the Era of Aero-Space Planes", by Russel J Hannigan, Krieger, 1994. Note 9 on page 183 discusses the cost per flight of the Space Shuttle.

19 "The Prospects for European Aerospace Transporters", by D M Ashford and P Q Collins, Aeronautical Journal of the R Ae Soc, Jan, Feb and March 1989.

20 "Prelude to Space Travel" by Wernher von Braun. Included in "Across the Space Frontier", edited by Cornelius Ryan. Sidgwick and Jackson Limited, London, 1952.

21 "Human flapping-wing flight under reduced gravity", by P Q Collins, Tokyo University, RCAST, and J M R Graham, Imperial College of Science, Technology and Medicine, Aeronautical Journal of the Royal Aeronautical Society, May 1994.

22 "Life at the Extremes, The Science of Survival", by Frances Ashcroft, Flamingo, 2000.

23 "Radiation Concern for Astronauts", by Erik Seedhouse, FBIS. Spaceflight, August 1999.

24 "The Cambridge Encyclopedia of Space", edited by Michael Rycroft, Cambridge University Press, 1990, page 143.

25 "The Design and Development of the Thiokol XLR99 Rocket Engine for the X-15 Aircraft", by Harold Davies, AFRAeS, Journal of the Royal Aeronautical Society, February 1963.

26 "Demand for Space Tourism in America and Japan, and its Implications for Future Space Activities", by P Collins, R Stockmans and M Maita. Presented at Sixth International Space Conference of Pacific-Basin Societies, Marina del Rey, California, December 1995.

27 "Handbook of Cost Engineering for Space Transportation Systems with Transcost 7.0" by Dr Dietrich E Koelle. Trans Cost Systems, D-85521 Ottobrun, Liebigweg 10, Germany, November 2000

28 "Tourism Cost Realities", Aviation Week & Space Technology, February 4, 2002, page 17.

29 "Operational Experience of the X-15 Airplane as a Reusable Vehicle System", by James E Love and William R Young, Presented to SAE Space Technology Conference, Palo Alto. California, May 1967. SAE Paper 670394.

30 "A French Concept for an Aerospace Transporter", by H Deplante and P A Perrier. Presented to SAE Space Technology Conference, Palo Alto. California, May 1967. SAE paper 670388.

31 "The Boeing TSTO", by R Hardy, L Eldrenkamp, and D Ruzicka. Boeing Defense and Space Group, Seattle, WA, USA. AIAA-93-5167. Presented at AIAA/DGLR Fifth International Aerospace Planes and Hypersonics Technologies Conference, Munich, December 1993.

32 "Small TSTO RLVs: Market-Building Stepping Stones to SSTO RLVs", by Len Cormier. AIAA-99-2620, presented at the 35th AIAA/ASME/SAE/ASEE Joint Propulsion Conference and Exhibit, June 1999, Los Angeles, California.

33 "A Preliminary Feasibility Study of the Spacecab Low-Cost Spaceplane and of the Spacecab Demonstrator", Bristol Spaceplanes Limited Report TR 6, February 1994. Carried out under European Space Agency Contract No. 10411/93/F/TB Volume 1 reproduced as "The Potential of Spaceplanes" in the Journal of Practical Applications in Space, Spring 1995. www.bristolspaceplanes.com

34 Market Research carried out by Harris Interactive and Space Adventures Ltd, 2001. info@spaceadventures.com. For more general information on space tourism, see www.spacefuture.com

35 "General Public Space Travel and Tourism" by Daniel O'Neil, Marshall Space Flight Center, Ivan Bekey and John Mankins, NASA Headquarters, and Thomas F Rogers and Eric W Stallmer, Space Transportation Association. NASA/STA joint study. NP-1998-03-11-MSFC, March 1998.

36 "Public Choice Economics and Space Policy: Realising Space Tourism", by Professor P Collins. IAA-00-IAA.1.3.03. Presented at 51st International Astronautical Congress, 1-6 Oct 2000, Rio de Janeiro, Brazil

37 House of Commons Trade and Industry Committee Tenth Report 'UK Space Policy', HC 335, 4 July 2000.

38 "The Overview Effect - Space Exploration and Human Evolution", by Frank White, Houghton Mifflin, 1987.

39 "Spirit of Apollo - A Collection of Reflective Interviews". Booklet published by the AIAA in July 1989 to commemorate the twentieth anniversary of the first landing on the Moon.

INDEX